North American Hummingbirds

North American Hummingbirds

AN IDENTIFICATION GUIDE

George C. West

University of New Mexico Press • Albuquerque

© 2015 by George C. West
All rights reserved. Published 2015
Printed in China
20 19 18 17 16 15   1 2 3 4 5 6

Library of Congress Cataloging-in-Publication Data
West, George C.
 North American hummingbirds : an identification guide /
George C. West.
    pages cm
 Includes bibliographical references and index.
 ISBN 978-0-8263-3767-2 (pbk. : alk. paper) —
 ISBN 978-0-8263-4561-5 (electronic) 1. Hummingbirds—
North America—Identification. I. Title.
 QL696.A558W473 2015
 598.7′64—dc23
                              2014046942

Cover art: *Violet-crowned Hummingbird in an Arizona Sycamore Tree* by George C. West.
Interior photographs and illustrations by George C. West unless otherwise noted.
Book Design by Catherine Leonardo
Composed in Bell Gothic Std.
Display is ITC Galliard Roman

This guide is dedicated to the memory of Allen J. Tozier, who spent many hours with me in the field and at the Madera Canyon banding station catching birds, taking photographs, and helping us process hundreds of hummingbirds. His friendship is greatly missed.

CONTENTS

ACKNOWLEDGMENTS ix

INTRODUCTION xiii

    How to Identify the Hummingbirds of North America xiv

    Scientific Names and Species Codes xx

LARGE HUMMINGBIRDS 1

    Magnificent Hummingbird 2

    Blue-throated Hummingbird 14

    Plain-capped Starthroat 24

MEDIUM-SIZED HUMMINGBIRDS 33

    Berylline Hummingbird 34

    Buff-bellied Hummingbird 43

    Violet-crowned Hummingbird 56

SMALL HUMMINGBIRDS 67

    White-eared Hummingbird 68

    Broad-billed Hummingbird 77

    Lucifer Hummingbird 89

    Ruby-throated Hummingbird 98

    Black-chinned Hummingbird 107

    Anna's Hummingbird 121

Costa's Hummingbird   134
Calliope Hummingbird   144
Broad-tailed Hummingbird   153
Rufous Hummingbird   167
Allen's Hummingbird   179

ACCIDENTALS   189
Green Violet-ear   190
Green-breasted Mango   193
Cinnamon Hummingbird   195
Xantus's Hummingbird   197
Antillean Crested Hummingbird   199
Cuban Emerald   201
Bahama Woodstar   203
Bumblebee Hummingbird   206

MEASUREMENTS AND WEIGHTS OF ADULT HUMMINGBIRDS   209

DIAGRAMS AND GLOSSARY OF HUMMINGBIRD CHARACTERISTICS AND TERMS   217

REFERENCES   225

INDEX   231

ACKNOWLEDGMENTS

Jesse and Retha Hendryx of Nogales, Arizona, shared their love of hummingbirds with the general public for many years. At their home, Ruth and Steve Russell banded hummingbirds during many migrations. Ruth graciously trained me to band hummingbirds with techniques very different from those involved in banding passerines, which I had been doing since 1960. With her help, I obtained federal and state permits for banding hummingbirds. I experimented with traps and bandmaking and set up stations at a number of local spots, finally settling on one of the best places in southeastern Arizona for a large variety and number of migrating hummingbirds: Tom and Edith Beatty's Miller Canyon Guest Ranch and Orchard, on the eastern slope of the Huachuca Mountains. There I met Sheri Williamson, who has been studying hummingbirds for many years. She gave me advice on many identifying techniques and helped with public education at the site. I was also fortunate to meet Susan Wethington, a recent doctoral student of Steve Russell's at the University of Arizona. She had studied the migration of hummingbirds, using banding as one of her tools, and she was anxious to continue her work after obtaining her degree. Together, we devised a protocol for monitoring populations of hummingbirds at a number of stations in Arizona. Simultaneously, Barbara Carlson from Riverside, California, was doing the same thing. The three of us formed the Hummingbird Monitoring Network (HMN) in 2002. The network now has some thirty sites from British Columbia to central Mexico, all operating under the same protocols of timing, number of traps, number of feeders, hours of trapping, and data collected, so that results can be statistically compared.

Acknowledgments

Ellen, my wife of thirty-five years, and I operated sites in Miller Canyon, Arivaca, and Madera Canyon; at our home in Green Valley; and at a few spots in Texas and New Mexico from 1999 to 2010. Altogether, we processed over 14,500 hummingbirds, including all of the regularly occurring species in Arizona. We thank the Beattys in Miller Canyon, Melva Robin in Arivaca, and Luis and Nancy Calvo at the Chuparosa Inn in Madera Canyon, Arizona, for hosting us every other week from March to October for many years. At these public sites, we met hundreds of people who were enthralled by seeing hummingbirds up close, learning about them as we processed each bird, and getting the chance to have hummingbirds placed on their open palms, where they would usually remain for a short time and then fly off. We had hundreds of dedicated volunteers to help us, chief of whom were Tom and Beverly Pickering in Miller Canyon, Bob and Georgia Puttock in Arivaca, and Bonnie and Al Tozier in Madera Canyon. Most of the in-hand photos in this guide were taken at these banding sites in Arizona and in Victoria, Texas, with Brent Ortego, who knows more about Buff-bellied Hummingbirds than anyone else. Most of the data in this guide about that species was gathered while working with him.

The methods of identifying ages and sexes of hummingbirds were developed over many years by hummingbird banders across the country. Steve and Ruth Russell published a hummingbird banders' guide, now available online and through the Bird Banding Laboratory in Patuxent, Maryland. Susan Wethington is working on a guide for banders in the western states. New ideas and techniques arise often, resulting in more accurate determinations of age and sex. We look forward to learning about them.

Photographs of birds in the hand were taken of living birds, which were released after processing. We did not have a professional setup to photograph the wings, tails, heads, and other features. We had to do the best we could by briefly interrupting the banding process to take photos quickly, so that we could feed the birds and get them on their way as fast as possible.

I appreciate the review of this guide by Susan Wethington, who brought me up to date on new techniques. Two other reviewers of this guide, Carol A. Butler, my coauthor of *Do Hummingbirds Hum?*, and Cindy Lippincott, who worked with me on *A Birder's Guide to Alaska*, were especially helpful with style and text suggestions. Professional birding guide Laurens Halsey contributed many photographs and gave suggestions for improving this guide for birders. Bruce Taubert allowed us to print some of his hummingbird portraits. Arizona birder Larry Liese assisted with the meanings of scientific names.

Printing of this guide was supported in part by a contribution from Bonnie Tozier, who, with her late husband Allen, volunteered assistance over many years to trap and process hummingbirds in Madera Canyon in the early mornings. The Hummingbird Monitoring Network—a science-based, project-driven nonprofit organization dedicated to the preservation of hummingbird diversity and abundance throughout the Americas—was instrumental, through Susan Wethington and her husband Lee Rogers, not only in designing equipment and providing supplies to banders, but also in requiring the necessary protocol to make our banding efforts meaningful. Contributions to support HMN's research may be sent to: Hummingbird Monitoring Network, PO Box 115, Patagonia, AZ 85624.

INTRODUCTION

This guide is designed for birders and banders to more easily and accurately identify, age, and sex the hummingbirds they encounter in North America north of Mexico. In addition to the seventeen species of hummingbirds that can be called dependable in their occurrence, there are eight others that have been seen here, although not with any regularity as to location or date. These are called Accidentals. Some of the accidental species have appeared many times and have been documented by specimen or photograph, while others have appeared only a few times, some long ago. Some have only been seen but not documented.

It is usually easier to identify adult male hummingbirds, if you are able to see the colorful gorget or crown. Identifying females and either sex of juvenile hummingbirds is usually much more difficult. Birders who cannot study a free-living bird "in the hand"—as a licensed hummingbird bander can—must rely on good optics, superior vision, and often high-speed, high-resolution photography to see the details of the bird's plumage. The use of ten-power binoculars or a spotting scope is sufficient to see most of the characters needed. It is often possible to approach birds at feeders, coming close enough to use a short telephoto lens and a high-resolution digital camera to "freeze" the bird with its tail feathers spread and capture the details needed. Cameras with over fifteen megapixel sensors that take photos saved at a printer resolution of at least three hundred dpi (dots per inch) should yield sharp enough images that they can be enlarged on the camera screen or computer monitor, showing most of the details a bander can see with the bird in hand. Some photographers use high-speed flash setups designed to stop all motion, but these are expensive as well as cumbersome for the average birder to transport and use.

HOW TO IDENTIFY THE HUMMINGBIRDS OF NORTH AMERICA

Experienced birders can often identify a bird by their initial impression upon first sighting. It takes time and experience to gain this ability. Hummingbirds are not always cooperative, and a first impression may be only a blur of color whizzing past you. But they do stop, perch, hover, and feed for extended periods—long enough for you to generate a good impression and see some details. Look for the bird's overall size; the shape and proportions of its body, including bill and tail; the gorget color and shape; the body color, both dorsal and ventral; the tail shape and color; the bill shape, color, and curvature, if any; the wing feather pattern; the bird's behavior; and its voice or any sounds it may make.

GENERAL SIZE: Measurements in this guide are given in millimeters (mm) and grams (g). You can convert metric to English measurements by knowing that 25.4 millimeters equals one inch and there are about 454 grams in one pound.

Is your first impression of a large hummingbird, a very small one, or one somewhere in between? There are roughly three sizes of hummingbirds found in North America: large, medium, and small. When measured from tip of bill to end of tail, the large birds are about 127.0–152.0 mm (5–6 inches) in total length, the medium birds 89.0–120.0 mm (3½–4¾ inches), and the small birds 64.0–89.0 mm (2½–3½ inches). In this guide, the species are listed by size: large (Magnificent Hummingbird, Blue-throated Hummingbird, and Plain-capped Starthroat), medium (Berylline, Buff-bellied, and Violet-crowned Hummingbirds), and small (White-eared, Broad-billed, Lucifer, Ruby-throated, Black-chinned, Anna's, Costa's, Calliope, Broad-tailed,

Rufous, and Allen's Hummingbirds). The Accidentals follow, unrelated to size (Green Violet-ear, Green-breasted Mango, Cinnamon Hummingbird, Xantus's Hummingbird, Antillean Crested Hummingbird, Cuban Emerald, Bahama Woodstar, and Bumblebee Hummingbird).

SHAPE AND PROPORTIONS: Is the bird long and sleek, short and dumpy, or just kind of normal? Is the bill very short or very long relative to the bird's body size? Is the tail short—extending only to the tip of the folded wing—or longer?

GORGET COLOR AND SHAPE: Does the bird have any color on the throat? Adult males of most species have some bright metallic color on the throat (gorget), and some females and juveniles may have many or only a few iridescent feathers in the gorget. Iridescence, and not only pigmentation, gives the gorget its color. The color you see depends on both the angle of the sun and your angle of observation in relation to the bird. What might appear brilliant red, green, blue, or violet when viewed with the bird facing into the sun may appear golden or black from another angle. While there is some pigment (melanin) in most of the hummingbird's colored feathers, the multiple layers of pigment and air-filled platelets in the tiny distal barbules that make up the feather refract incoming light and shine back with certain colors, depending on the composition of the layers of platelets. Does the gorget just fill the throat, or are there long "ear" feathers that point backward? Is the color restricted to just the center of the throat, or are there only a few scattered feathers there? Are the scattered feathers large or small? Are they dull or very shiny?

BODY COLOR: What is the color of the back and undersides? As almost all hummingbirds are green on the back, that character alone is of little help. But some are browner, grayer, or rufous (rusty) in color. The color of the undersides (breast, belly, and vent) should be noted. The background of the undersides may be

Introduction

First impression: a small hummingbird with a relatively short, straight bill; chunky body; short tail; and purple gorget and crown. It must be an adult male Costa's Hummingbird.

white to gray to buffy to black in color (in good light), with or without spots. The flanks may be clear, rusty, or very "dirty" looking. Juvenile hummingbirds of all species that breed in North America have individual feathers on the crown and back tipped with a lighter color. Banders often call this appearance "buff," "scaly," or "buffy" back. This lighter color—usually buffy, cinnamon, or dull white—gives the crown and back a "scaly" appearance. The pale fringes of these feathers wear off gradually, so that by fall, most birds hatched in early spring have lost their scaly backs.

TAIL SHAPE AND COLOR: Is the tail squared off at the end, rounded, slightly notched, or deeply forked? Are the individual feathers rounded or pointed at the tips? Are some feathers much narrower than others? The distribution pattern of colors on individual tail feathers, the amount of white at the tips of outer tail feathers, and the relative width of tail feathers can be used to distinguish ages and sexes for many species.

Introduction

Adult male Anna's Hummingbird, showing the variation in color of its bright rose-pink crown and gorget, due to the angle of light.

BILL SHAPE AND COLOR: Is the bill short and straight, proportionately long, or downcurved? Some birds have very straight and relatively short bills, others have very long bills that are only slightly downcurved (decurved), and some species have markedly decurved bills. Most hummingbirds have black bills, but some species have a reddish or pinkish color, especially along the base of the lower mandible. Others have a bright red bill with a black tip.

The most reliable characteristic to separate young from adult birds is the presence of bill corrugations. These are fine, oblique striations along the lateral surface of the maxilla (upper bill), which are present in young hummingbirds for several months after fledging. Banders refer to the corrugations as "grooves." They are most evident right after fledging, and they decrease in depth and extent with age and as the bill hardens; when the bill hardens, they disappear altogether. Although you can see the grooves of a very young bird's bill with the naked eye—especially on a large hummingbird in good light—magnification (through the use of a lens) of about ten times is usually required to see them

Introduction

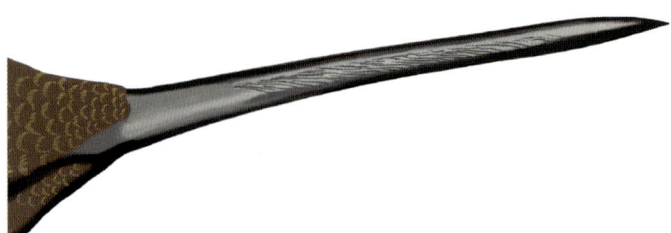

Bill of a juvenile Black-chinned Hummingbird, showing the corrugations or "grooves" that indicate that the bird is still in its first year of life.

clearly enough to estimate their extent. It is possible to see the grooves of a young bird using a good spotting scope.

WING FEATHER PATTERN: Is there any color (other than gray) in the wings? Is the shape of the wing feather tips anything but rounded? The wings of most hummingbirds are uniformly greenish to gray above, with uniformly gray flight feathers (primaries and secondaries). One of our species, the Berylline Hummingbird, has a distinct cinnamon patch on the wing. Older feathers in the wing can be distinguished from new or recently molted feathers by texture, wear, and intensity of color, with new feathers being darker, clean-edged at the tips, and dense throughout. Molt of the flight feathers occurs at different times for different species. The relative size of the vanes on each side of the primary feathers' shafts can distinguish Black-chinned and Ruby-throated Hummingbirds from Anna's and Costa's Hummingbirds. With the bird at rest, you can see this difference using binoculars. You can also look for the shape of the tips of the flight feathers—individual feathers may be pointed, rounded at the end, squared off, notched, or angled.

BEHAVIOR: Is the bird very aggressive and, despite its small size, does it try to drive larger birds away from its favorite feeder? Or is it more submissive, and does it wait its turn? The smaller species have very fast wing beats that produce the familiar humming sound, while the larger ones have slower wing beats, and the sound is more of a whirr than a hum. Some species have specialized wing feathers that make a zinging or ringing sound in forward flight. Some make loud chirping noises with their tails during courtship flight. Some birds constantly flash their tails by spreading them open when confronting another bird; some perch with the tail completely closed, so that it becomes very narrow. Some hover with the tail stationary or moving only slightly, and some hover with it pumping up and down continuously. Most males perch in a conspicuous spot from which they can overlook their territory, moving their heads left and right and often calling at the same time. You will notice other characteristic behaviors as you learn more about each individual species.

VOICE: Is the bird silent or constantly "talking"? The chattering notes of birds around a feeder are useful in recognizing species, as are the songs or calls of courting and territorial males. Textual descriptions of hummingbird vocalizations are typically inadequate. Although some vocalization characteristics are included in the following species accounts, the reader is strongly encouraged to listen to and learn hummingbird vocalizations from visually identified birds and from recordings (see References for suggestions).

SUMMARY: Many of the characters are obviously much better seen with the bird in hand. Banders have the opportunity to make measurements; check for corrugations on the bill; and determine the amount of fat, state of reproduction, stages of molt, and all of the plumage characteristics, in order to confirm species and make age and sex determinations. Measurements of wing and culmen are helpful, and the width of the outer rectrix of many hummingbirds is essential to identifying adult from juvenile females and to differentiating some Allen's, Rufous, and Broad-tailed Hummingbirds. Measurements and weights are given in a later chapter in this guide (see Measurements and Weights of Adult Hummingbirds).

Introduction

SCIENTIFIC NAMES AND SPECIES CODES

Scientific names are universally accepted and are used by birders and ornithologists throughout the world. The name may change only after research shows that there was an earlier error, or when new information comes to light through genetic research on the relationships of species. Common names vary in different parts of the world, but with increased international communication, even the common names may eventually become standardized.

The scientific names show the relationships among species, for instance, whether or not species are in the same genus. All hummingbirds are in the same family, *Trochilidae*. In the accounts of each species, the scientific name—which is usually derived from Greek or Latin—is followed by the most frequently used Mexican common name, then the meaning of the scientific name. The author of each name (when known) is in parentheses following the meaning of the name.

The federal Bird Banding Laboratory developed a four-letter code for all North American birds, which banders use in their reports. Most of the codes are the first two letters of the first word in the common name and the first two letters of the last word, e.g., **RUHU** = **Ru**fous **Hu**mmingbird. However, there are many birds with more than a two-word common name, and some species have common names that would result in duplicate codes. These variations have been considered, and there are no duplicate codes. More birders are learning the codes, as they are faster to write down in the field and shorter in text communications. The four-letter codes are included for each of the species in this guide.

LARGE HUMMINGBIRDS

Birds with a total length of five inches or more, including the bill

Large Hummingbirds

MAGNIFICENT HUMMINGBIRD, MAHU (*Eugenes fulgens*)
Colibrí Magnífico (Sp.) *Eugenes* = attractive in appearance
(Gould); *fulgens* = glittering or shining (Swainson)

IDENTIFICATION: The Magnificent Hummingbird is one of three large hummingbirds that occur regularly in North America. The bill is long and the head looks proportionately large for the body. The crown, back, and tail are bronzy green. The undersides are dark in males and light in females and juveniles. The forehead feathers extend over the nares (nostrils).

ADULT MALE: Bright green to turquoise blue-green gorget and violet crown. The crown and gorget often appear black, unless the bird is facing you. Combined with their very dark undersides, this makes males look very dark. The notched tail is all bronzy green. Adults even several years old may not have complete violet crowns forward to the base of the bill.

ADULT FEMALE: The face has a white postocular spot and a light line back of the eye. The undersides are pale, with grayish-green

Adult male.

Magnificent Hummingbird

Adult male Magnificent Hummingbird, dorsal view.

Adult male, ventral view.

Adult female, dorsal view.

Adult female, ventral view.

Large Hummingbirds

Juvenile male, dorsal view.

Juvenile male, ventral view.

Juvenile female, dorsal view.

Juvenile female, ventral view.

Magnificent Hummingbird

Adult male tail. Note the all-bronze tail, except for pale edges at the feather tips.

Adult female tail. Note the green of the central tail feathers (r1) and black of the other tail feathers, with white tips. Compare with the Blue-throated Hummingbird adult female tail, which has blue-black central tail feathers.

Large Hummingbirds

Juvenile male tail. Note the pale gray tips on r4 and r5, with thin pale edges to the tips of r2 and r3. Only r1 is similar to the adult tail.

Juvenile female tail. Note the green central feathers and large white tips on r3, r4, and r5. There is usually also some white at the tip of r2. This pattern and the overall coloring are close to that of the adult female. Compare with the Blue-throated juvenile female, whose central tail feathers are blue black. Its other feathers have even larger white terminal patches.

Magnificent Hummingbird

Older male's tail, one or more years after hatching. Since about 10 percent of second-year and after-second-year males retain white tips to r3, r4, or r5, we can only say that this bird was not hatched in the year the photo was taken. In order to determine that a male is in its first year after hatching, i.e., its second year, we would need to see a partially developed gorget and crown, white or gray at the tip of one of the outer tail feathers (r3–r5), and significant wear to the tail feather tips.

Adult male. The bird is very dark, with no sign of violet on the crown and only a few green feathers visible in the gorget.

Large Hummingbirds

Adult male. The crown and gorget appear black, as the bird is not facing the light. Note that there are no white spots on the tail feather tips.

This adult male looks almost all black when perched out of the light.

Magnificent Hummingbird

Juvenile male. The gorget is just developing, but the rest of the undersides look like those of a female, without any dark green feathers.

Juvenile male. Note the few violet feathers over the eyes and the scaly appearance to the crown.

Large Hummingbirds

Adult female. Note the dark spotting on the whitish throat. Compare with the adult female Blue-throated Hummingbird, which is a more washed-out gray. The white head stripe of the Blue-throated is more over the eye, longer, and better defined. Check the color of the central tail feathers: if they are greenish, it is a Magnificent, and if they are blue black, it is a Blue-throated.

Older male, one or more years after it was hatched. The photo was taken in April. Note that the green spangles do not fill the gorget, and the purple spangles on the crown are not complete, even on the top of the crown. Neither area may ever be complete, even in older birds.

Another view of the same bird as in the previous image.

Magnificent Hummingbird

Juvenile female. Note the scaly appearance of the crown and back feathers. The white postocular stripe is not as definitive and prominent as the stripe of a Blue-throated Hummingbird. The throat, breast, and flanks are heavily spotted with bronzy feathers, unlike the fuzzier gray undersides of the Blue-throated.

Adult female in the field, showing a light postocular stripe that is not as prominent as that of a female Blue-throated. Also note the spotted undersides, unlike those of the female Blue-throated.

Large Hummingbirds

Adult male in flight, showing dark undersides, a green gorget, and a violet crown. The tail appears all black from this angle, but it is bronze colored when viewed from above.

spotting on the throat. There are gray spots and streaking on the breast and belly, as well as conspicuous white tips on the bronzy-green and black tail feathers r3 through r5, with varying amounts of white on r2.

JUVENILES: The bill has grooves or corrugations on the maxilla. Each crown and back feather has a conspicuous whitish tip.

JUVENILE MALE: There may be some green spangles on the throat in very young birds, which increase in number over the summer. There should be a few violet feathers over the eyes; these will eventually cover much of the posterior crown. In some individuals, the anterior crown may never become filled with iridescent violet spangles. The tail is bronzy green, with a thin light tip to each feather.

JUVENILE FEMALE: Resembles the adult female. There may be more white at the tips of tail feathers r2 through r5 than in the adult female tail.

Magnificent Hummingbird

VOICE: Call notes are sharp *chick* or *tzick* notes. The song of a perched male is long and bubbly.

SIMILAR SPECIES: The Magnificent tail is bronzy at the base, and the central pair of tail feathers is green and not solid blue black—as in the Blue-throated. The back is usually more bronze green than that of the grayer Blue-throated. Although the female Magnificent has a light line back of the eye, it is not as prominent as that of the female Blue-throated. Note the color and shape of the tail: it is all blue black and rounded, with large white tips, in the female Blue-throated, while it is partly bronze and black and slightly notched, with small white tips, in the female Magnificent.

DISTRIBUTION: Magnificent Hummingbirds range from southeastern Arizona and southwestern New Mexico to the Davis Mountains in Texas through Sierra Madre Occidental south, intermittently to Panama. On rare occasions, they wander into southern New Mexico, southwestern Colorado, and southern California. A population extends down the higher elevations of the Sierra Madre Oriental to connect with the western group in central Mexico. Some of the population that breeds in the United States, northern Mexico, and parts of central Mexico migrate south and winter in Mexico, while the rest of the population is resident.

MIGRATION: Northward-migrating males arrive in southeastern Arizona in March or early April. Many birds are resident and move down the mountain slopes to lower elevations in winter, and then back up again in spring. Southward-departing birds may leave as early as late June in southeastern Arizona. Some Magnificents wander into the White Mountains northeast of Phoenix, Arizona, in late summer, and they may be expanding their range northward.

COURTSHIP AND NESTING: Magnificents prefer medium- to high-elevation riparian habitats in the sky islands of southeastern

Arizona, where oak (*Quercus* sp.), juniper (*Juniperus* sp.), sycamore (*Platanus* sp.), walnut (*Juglans* sp.), fir (*Abies* sp.), Douglas fir (*Pseudotsuga menziesii*), and pine (*Pinus* sp.) forests predominate. In good years, a second brood will carry the nesting season into August. Young are seen from May through September, when most birds depart Arizona. Nests are usually constructed on a horizontal branch, which often overhangs a stream from three to twenty-seven meters above the ground and two to three meters out from the tree trunk. The female incubates two eggs for only fifteen to nineteen days, and fledging occurs around twenty-five days after hatching. There is little data on either of these time intervals.

NUTRITION AND MOLT: Unlike most hummingbirds, large Magnificents rarely interfere with the feeding of neighboring Black-chinned or Broad-tailed Hummingbirds, or that of migrating Rufous Hummingbirds. They are displaced by the slightly larger Blue-throated Hummingbirds. The little data that exists to show their preference for nectar-producing plants indicates that they choose century plant (*Agave* sp.), bouvardia (*Bouvardia* sp.), columbine (*Aquilegia* sp.), hedgehog cactus (*Echinocereus* sp.), and penstemon (*Penstemon* sp.), among other flowers. Their long bills allow them to extract sucrose solutions from flowers with long corollas. They also spend a lot of time fly-catching insects and gleaning them from plants. Magnificents can fly many miles in a single day as they "trapline" feed on their own range, and they often cross over fifty miles to other sky islands to feed. Adults molt only once a year, but the molt may be divided into two seasons: late winter and early spring, and mid- to late summer. More study is required, as a complete molt has not been documented in the United States.

BLUE-THROATED HUMMINGBIRD, BLUH (*Lampornis clemenciae*) Colibrí-serrano Gorjiazul (Sp.) *Lampornis* = lamp or torch (shining) bird (Swainson); *clemenciae* = named for Clemence Lesson, wife of the author (Lesson)

Blue-throated Hummingbird

The adult male is unmistakable when viewed in good light. Note the gray wash to the undersides (unlike the Magnificent) and the large white spot at the tip of r5, visible between the wing tips.

IDENTIFICATION: The Blue-throated is the largest hummingbird in North America, only a few millimeters larger than the Magnificent Hummingbird. All ages and sexes have an olive- to dusky-green crown and back, darkening toward the tail; a gray breast and belly; and an all-dark blue-black tail with large white tips. The forehead feathers do not cover the nares.

ADULT MALE: The throat is dull to bright blue in good light, and it often appears dull gray. The tail has a little less white than that of the female.

ADULT FEMALE: There are no blue feathers on the gray throat. There is a conspicuous, long white postocular stripe.

JUVENILES: Resemble females. There are light fringes to each crown and back feather and grooves on the maxilla until late summer.

Large Hummingbirds

JUVENILE MALE: Males at hatching have some blue spangles on the throat, and they otherwise resemble adult females.

JUVENILE FEMALE: There are no blue spangles on the throat. After the scaly back and grooves are gone, it is not possible to reliably distinguish between juvenile and adult females.

VOICE: This species has a wide variety of calls and songs, the most familiar being the persistent *seep* call of males on territory.

SIMILAR SPECIES: The similarly sized female Magnificent Hummingbird has a bronzy-green tail, usually with much smaller white tips. If the central tail feathers of a large hummingbird are blue black, it is a Blue-throated. If the central tail feathers are green, it is a Magnificent. The postocular white stripe of the female Magnificent is smaller than and not as sharply defined as that of the female Blue-throated, but the central tail feather color is a better character if you are not sure of the eye stripe. The forehead feathers of Magnificent Hummingbirds cover the nares.

DISTRIBUTION: The Blue-throated Hummingbird's breeding range is in the sky islands of southeastern Arizona and the northern Sierra Madre Occidental of Sonora, Mexico. There are populations in Big Bend National Park, Texas, and several in New Mexico. Wandering birds have been sighted south of Flagstaff, Arizona; in western California; in the Rocky Mountains of Colorado; and in scattered locations in Texas, east to Houston and Brownsville. They prefer high-elevation pine (*Pinus* sp.) and fir (*Abies* sp.) forests in Mexico and middle elevations along perennial streams in the canyons of southeastern Arizona, southwestern New Mexico, and southwestern Texas.

MIGRATION: While some individuals remain in Arizona all winter, most depart for warmer temperatures in Mexico and return the next spring, in March or April. The migration of some of the

Blue-throated Hummingbird

Adult male Blue-throated Hummingbird, dorsal view.

Adult male, ventral view.

Adult female, dorsal view.

Adult female, ventral view.

Large Hummingbirds

Juvenile male, dorsal view.

Juvenile male, ventral view.

Juvenile female, dorsal view.

Juvenile female, ventral view.

Blue-throated Hummingbird

Adult male tail, showing prominent white at the tips of only r4 and r5.

Tail of an adult female, showing the dark tail, with large white tips on r5 and r4 and some white on r3. Compare with the female Magnificent Hummingbird, whose central pair of tail feathers is green.

Large Hummingbirds

Tail of a juvenile male, showing the all-dark blue-black tail, with large white tips on r5 and r4 and some white on r3. Compare with the female Magnificent Hummingbird.

Juvenile female tail, showing large white tail tips on r4 and r5 and smaller tips on r3 and often r2.

Blue-throated Hummingbird

Juvenile female Blue-throated Hummingbird. Note the long white postocular stripe, which curves down behind the eye and the more evenly gray throat. Compare with the Magnificent adult and juvenile female.

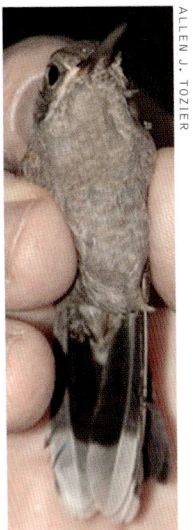

Adult female, showing its gray belly, large white tail tips, and lack of blue spangles on the throat.

Juvenile male with a partly developed gorget.

Large Hummingbirds

Juvenile male.

Juvenile female. Note the pale fringes to the crown and back feathers, called "buff back" by banders. The long, well-defined white head stripe that starts over the upper rear corner of the eye is typical of Blue-throated Hummingbirds.

Blue-throated Hummingbird

Juvenile female. The "buff back" is not as clearly seen on this bird, but the long white eye stripe and malar stripe are clearly visible.

Juvenile female. This is the same bird as in the previous photograph.

Adult male in flight. Note the all-gray plumage except for the blue gorget, white postocular stripe, and black tail with white tips on the outer two pairs of feathers.

US population is probably altitudinal, but most birds migrate south. Mexican birds have been found at lower elevations in drier habitats in winter. Those birds that migrate south leave for the winter between September and November.

COURTSHIP AND NESTING: Females start nest building in early April in Arizona, in locations that are usually covered, such as under stream bank cuts, rocks, cave entrances, or man-made structures (projections under the eaves of houses, porches, light fixtures, etc.). The female lays one to two eggs, usually within forty-eight hours, and incubates for seventeen to nineteen days. Young are fledged twenty-four to twenty-six days after hatching. All care of the young is by the female. Some females may have three broods per year.

NUTRITION AND MOLT: The Blue-throated is a very aggressive bird, capable of dominating feeders by driving away all other species. Blue-throateds feed on a large variety of insects caught in flight and on larger stems, branches, and tree trunks. They also feed on nectar from flowers such as penstemon (*Penstemon* sp.), sage (*Salvia* sp.), century plant (*Agave* sp.), columbine (*Aquilegia* sp.), locust (*Robinia* sp.), and cardinal flower (*Lobelia cardinalis*). They are less dependent in summer on artificial feeders than most other species, although they rely on them for their primary winter energy source. We are unsure when flight feather molt begins, but many birds banded in Arizona are in primary feather molt. Body molt must occur in winter before spring migration.

PLAIN-CAPPED STARTHROAT, PCST (*Heliomaster constantii*) Picolargo Coronioscuro (Sp.) *Heliomaster* = sun breast; *constantii* = unknown person's name

IDENTIFICATION: The Starthroat is very rare, but it is becoming more regularly seen in southeastern Arizona. Starthroats have very long, straight bills. The crown, back, and most of the dorsal tail feathers are a dull bronzy green, which usually appears olive

Plain-capped Starthroat

Adult Plain-capped Starthroat. Note the long bill, white postocular stripe, white malar stripe, and relatively inconspicuous red spangles at the bottom of the gorget.

brown to dark brown. There are wide pale postocular and malar stripes, and the undersides are grayish with a paler central vertical stripe. A bright white flank is seen only in flight or when the bird is preening. There is a white patch on the lower back, which is often hidden, as well as a white puffy vent band. The wings extend beyond the short tail. The bronzy tail is almost square at the tip; it is blackish toward the end of each feather. There are white spots on the inner web tips of r2 through r4 and larger white tips on r5. No white appears on the central pair of tail feathers.

ADULTS: Sexes are similar. The gorget is dark ruby red, but the red spangles are mostly seen at the lower end of the gorget. In some birds, each gorget feather may have a buffy or whitish margin. Females usually have less red in the gorget.

JUVENILES: Similar to adults, except that the gorget usually lacks red feathers or has only a few by late summer. The gorget is very dark; each feather has a pale margin. The bill has corrugations for several weeks post-hatching. The feathers of the crown and back show a scaly appearance, as each feather has a pale terminal fringe. The obvious malar and postocular stripes are dull white or even buffy.

SIMILAR SPECIES: The Starthroat has a longer bill than either the Magnificent or Blue-throated Hummingbirds. The body weight is about the same as that of the Magnificent Hummingbird.

DISTRIBUTION: The Starthroat is resident in western Mexico—from southern Sonora south to Honduras—in arid to semiarid woodlands and forest edge; along streams; and in pine/oak woodlands in mountain canyons. It occurs almost annually in southeastern Arizona, primarily in the sky island habitats in the Chiricahua, Huachuca, and Santa Rita Mountains. However, it has been seen as far north as Phoenix, Arizona, and, rarely, in southwestern New Mexico.

MIGRATION: The Starthroat is resident over most of its range from northern Sinaloa southward, but it disperses after breeding to the north into Sonora and Arizona. Some individuals spend the summer and fall in southeastern Arizona.

COURTSHIP AND NESTING: The Starthroat is not known to have nested north of Mexico as of 2014. In Mexico, courtship consists of a display in which the male makes short vertical flights in front of the female. Nests are built high in trees and made of plant fiber, plant down, and grasses woven together with spider webbing. Bits of bark, lichen, and leaves may be added to the outside for camouflage.

NUTRITION: Starthroats feed primarily on insects and attend nectar flower patches and feeders. Some individuals have been known to show up at the same feeder daily throughout most of the summer, at the same time.

Plain-capped Starthroat

Adult Plain-capped Starthroat, dorsal view.

Adult, ventral view.

Juvenile, dorsal view.

Juvenile, ventral view.

Large Hummingbirds

Note the white postocular stripe, white malar stripe, white on the tail tips, and white on the lower back.

This adult Starthroat conveniently raised its wing to show the bright white flanks. The white patch on the lower midback is also evident.

Plain-capped Starthroat

Juvenile showing buffy fringes to the back feathers. The flight feathers are being molted, with some of the inner ones already grown in. The thin, transparent juvenile feathers are easy to see.

This adult shows off what few red gorget feathers it has, as well as gray flanks, a white midbelly, and the unusual pattern of white and black on the tips of the tail feathers. Note that the r2, r3, and r4 feather tips are white only on the medial half of the feather, and the lateral side is black.

Large Hummingbirds

Comparison between a Plain-capped Starthroat (*left*) and an adult male Black-chinned Hummingbird, showing their size difference.

This bird appears to be molting its crown feathers, although it is May. The tail tip shows no white spots when it is closed, because r1 has no white tip.

Plain-capped Starthroat

A flying adult Plain-capped Starthroat is almost twice the size of the female Black-chinned Hummingbird at the feeder. With its long bill and head pattern, the bird is hard to mistake for any other North American hummingbird.

MEDIUM-SIZED HUMMINGBIRDS

Birds with a total length of about 4 to 4½ inches, including the bill

Medium-Sized Hummingbirds

BERYLLINE HUMMINGBIRD, BEHU (*Amazilia beryllina*)
Colibrí de Berilo (Sp.) *Amazilia* = word of unknown origin; *beryllina* = green colored

IDENTIFICATION: Berylline Hummingbirds are bright to dusky green, with a cinnamon or orange-brown wing patch that extends from the secondaries into the primaries. The tail is dark rufous, and in good light there is a purplish iridescence to many of the feathers. The maxilla is black, and the mandible is often dull red.

ADULT MALE: The head and back are bright green. The gorget, breast, and upper belly shine brilliant green in the sun. The lower belly is graybuff to buffy to cinnamon, with a white area around the vent. The undertail coverts are cinnamon. There is a small white postocular spot. The rufous wing band present in all ages and sexes is easily seen in flight, as the underwing coverts are also cinnamon. The slightly notched tail usually appears dark, but it shows a rusty color with a purple shine in good light.

Adult Berylline Hummingbird, showing the diagnostic cinnamon color on the secondary and inner primary feathers and iridescent violet of the tail.

Berylline Hummingbird

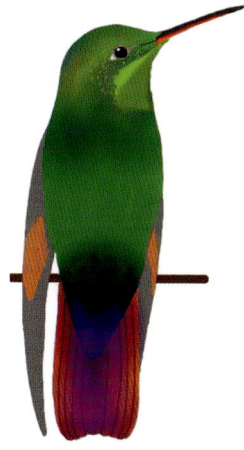

Adult male, dorsal view.

Adult male, ventral view.

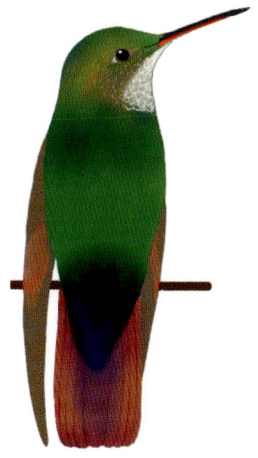

Adult female, dorsal view.

Adult female, ventral view.

Medium-Sized Hummingbirds

Juvenile male, dorsal view.

Juvenile male, ventral view.

Juvenile female, dorsal view.

Juvenile female, ventral view.

Berylline Hummingbird

Adult male tail. The tail shows its reddish color with a bright purple iridescence.

Adult female tail. The tail shape and coloration are very similar to those of the male.

Medium-Sized Hummingbirds

Adult male wing. The wing of the male shows more rusty orange than that of the female, extending into the ninth primary (p9) and almost to the tip of p1 and the first few secondary wing feathers.

Adult female wing. The wing shape and coloration are also similar to those of the male, but the extent of rusty orange is less than in the male's wing. The underwing rusty cinnamon color is also slightly subdued in the female, although it is still very evident.

Berylline Hummingbird

The underside of the mandible of this male does not show any red. The gorget is packed with spangles, which shine much more brilliantly than the female's gorget.

This is the same bird as in the previous image, from a different angle. Now the gorget is a much darker green and has a hint of dark blue. Note that the belly is cinnamon buff.

In good light, it is easy to see the solid-green head, neck, gorget, and breast of an adult male. Note the bright cinnamon underwings and the cinnamon on the undertail coverts.

Medium-Sized Hummingbirds

Adult female. Note that the spangles on the gorget are greatly subdued, are spaced farther apart or are smaller, and do not give off the bright green shine of the male's gorget.

The underside of the mandible is reddish in this adult female Berylline. The rusty color on the wing is very evident in both sexes.

Berylline Hummingbird

Adult male, showing the dark green breast and gorget and obvious cinnamon in the wing.

ADULT FEMALE: Similar to the male, although the gorget, breast, and upper belly are not solid green; they are broken up to show a pale background, especially on the chin and upper belly. The rufous wing patch is less extensive. The tail is square and rusty, with less purple shine than on the male.

JUVENILES: Similar to adult females, except that each crown and back feather has a pale cinnamon fringe. The pattern of green on the undersides is more diffuse and ragged than on the adult. The sexes are very difficult to distinguish in the field.

JUVENILE MALE: The gorget, breast, and upper belly have scattered groups of bright green iridescent feathers, which show more shine or iridescence than juvenile or adult females. The rufous wing patch is brighter and more extensive than that of either adult or juvenile females.

JUVENILE FEMALE: Like the juvenile male, except that the groups of green feathers on the undersides are not as bright or metallic looking. The rusty wing patch is less evident than in males.

SIMILAR SPECIES: In poor light, a Berylline seen from the back might be mistaken for a Buff-bellied or Cinnamon Hummingbird. However, both lack the cinnamon wing patch and have a rufous tail with no purple shine. From the front, the Cinnamon has no green gorget, and it is all cinnamon below.

DISTRIBUTION: The Berylline is primarily a Mexican bird that only reaches the United States with regularity in southeastern Arizona. There are records for Texas's Davis Mountains and Big Bend area, and for the southwestern corner of New Mexico. It is resident from southeastern Sonora and southwestern Chihuahua south through the Sierra Madre Occidental, and through southern Mexico into Honduras. Beryllines prefer deciduous forests and forest edge, woodlands, parks, and riparian areas, especially areas dominated by oaks (*Quercus* sp.). In Arizona, they are almost always found in mountain canyons, such as in the eastern slope of the Huachuca Mountains and, more rarely, in the Chiricahua and Santa Rita Mountains.

MIGRATION: Except for the northern breeding populations, the species is resident. Some may retreat from higher elevations in the cold of winter. Birds are usually first seen in Arizona in May or June, and they depart in early fall.

COURTSHIP AND NESTING: Courtship behavior has been little studied. Males set up territories and sing from perches. Several nests have been seen in Arizona, where they were built twenty to thirty feet above ground in small branches of the Arizona sycamore (*Platanus wrightii*). They are typical hummingbird nests, made of bits of thin leaves, plant fibers, and grasses woven together with spider webbing, and decorated with light-colored pieces of bark and lichen.

NUTRITION AND MOLT: Beryllines prefer insects that they hawk from perches (like flycatchers) and glean from foliage. They flit from flower to flower to feed on nectar, usually dominating smaller hummingbirds but in turn often being dominated by the more aggressive Blue-throated. There is no information on the progression or timing of molt of Arizona birds, but in Mexico, primary molt begins in spring and continues through fall.

BUFF-BELLIED HUMMINGBIRD, BUFH (*Amazilia yucatanensis*) Colibrí Vientre-canelo (Sp.) *Amazilia* = word of unknown origin; *yucatanensis* = from the Yucatán Peninsula, Mexico

The Buff-bellied is a medium-sized hummingbird in the genus *Amazilia*, which also includes the Berylline, Violet-crowned, and Cinnamon. Only the Buff-bellied, Violet-crowned, and Berylline are regularly occurring species in North America. The Cinnamon is a very rare accidental species with only two records in the United States, the last in 1993.

IDENTIFICATION: The Buff-bellied is a green hummingbird with a pale, rust-colored belly, a red bill, and a notched rusty tail.

ADULT MALE: The crown and back are bright yellow green to bronzy green. The culmen of the maxilla is bright blood red with a black tip. The underside of the mandible is also red. The gorget and breast are bright apple to emerald green. Iridescent green feathers may run up to and behind the eye. The center of the gorget often appears dark blue. Each spangle is relatively large, without pale fringes at the tip. The chin is covered with spangles and is not pale. The belly is pale cinnamon or buffy rust; the vent band is white. The undertail coverts are cinnamon, and the upper tail coverts are tipped with rusty red brown. The tail is rusty brown. The central pair of rectrices (r1) is usually darker rusty red brown, although sometimes it is greenish bronze. The outermost rectrix (r5) and sometimes the fourth rectrix (r4) may lack brown tips, but the second (r2) and third (r3) rectrices usually

Medium-Sized Hummingbirds

Adult male. Note the rusty upper tail coverts and cinnamon belly.

have dark olive-brown tips. There is considerable variation in the pattern of brown in the tail. The tail is notched 5.0 mm or more, measured from the longest rectrix (r4) to the shortest, the central pair (r1).

ADULT FEMALE: The female is a paler version of the male. The crown and back are yellow green to olive green. The culmen of the maxilla is dusky blood red with a diffuse black tip. The underside of the mandible is pink to red. The bright green iridescence of the gorget is confined to feathers mostly around the periphery of the gorget, instead of the center. Each feather is smaller than the spangles of the male. Many have pale marginal fringes, especially in the upper gorget. The chin is usually without green spangles. The lower breast and belly are very pale cinnamon, sometimes appearing without any rusty brown color. The lower belly is pale,

Buff-bellied Hummingbird

Adult male Buff-bellied Hummingbird, dorsal view.

Adult male, ventral view.

Adult female, dorsal view.

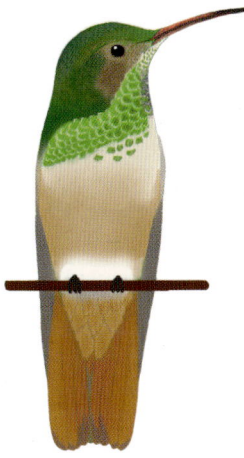

Adult female, ventral view.

Medium-Sized Hummingbirds

Adult male tail, showing that all feathers are reddish brown, with r1 being darker. The tips and margins of r1 to r5 have some dark brown.

Tail of an adult male (with only one r1 visible), showing r1 with some greenish color.

Buff-bellied Hummingbird

Adult female tail, showing that r1 is a darker bronze brown.

A different adult female tail, showing that r1 is bronze green. There is much variation both in the dark brown color in the tail and in the whole tail when it is viewed from different light angles.

Medium-Sized Hummingbirds

The juvenile male has larger spangles and some chin spots.

The juvenile female lacks chin spots and has smaller spangles on the gorget.

Buff-bellied Hummingbird

This adult male shows the cinnamon buff belly, darker cinnamon undertail coverts, and white vent band. Note the large green lower breast spangles.

This adult female also shows the cinnamon buff belly, darker cinnamon undertail coverts, and white vent band. The large green lower breast spangles are lacking in the female.

Medium-Sized Hummingbirds

Adult male. Note the full gorget to the base of the bill and the blue spangles from this light angle.

Adult male. The full gorget is brilliant green from another light angle.

Buff-bellied Hummingbird

Adult male bill. The bill is bright red, and the black on the tip is very short.

Adult female bill. The bill is duller red than the male's bill. The black at the tip is not as well defined, and it runs farther up the bill

A juvenile male, showing the lack of large spangles on the gorget.

A juvenile male, showing its black maxilla.

Medium-Sized Hummingbirds

Adult male in the field.

Adult female in the field. Note the lack of green spots on the breast and the dark tips of the rectrices.

the vent band is white, and the undertail coverts are pale cinnamon. The upper tail coverts are tipped with cinnamon but are more bronzy green than in males. The central pair of tail feathers is usually bronzy green mixed with rust. The tail often has more dark olive brown than the male. The tail notch is less than 5.0 mm deep.

JUVENILES: Grooves in the maxilla remain for several weeks after fledging. Since adults start nesting as early as February in the United States, one should not rely on the presence of corrugations to determine age later in the summer and fall. The crown and back feathers have pale rusty tips for several weeks after fledging. The plumage is slightly duller than that of the adult female. The maxilla is entirely black when birds are young, and it gradually becomes red from the first spring through winter. Distinguishing the sex of juveniles is difficult and usually requires a measurement of the depth of the tail notch (5.0 mm or more is probably a male).

JUVENILE MALE: Some bright green spangles grow in the throat, along with duller feathers. The center of the gorget is without iridescence and often appears dark blue. The chin area usually has some dull green spangles with buffy fringes.

JUVENILE FEMALE: Resembles a duller adult female, with bill corrugations and pale cinnamon fringes to the crown and back feathers. The gorget has smaller and more scattered iridescent green feathers than those of the juvenile male. The chin typically lacks spangles. The tail notch is less than 5.0 mm once the tail is completely grown in.

VOICE: The commonest calls are a *tic* or *tchik* note, called a "click" by some, while feeding or in response to a "mewing" call. Buff-bellieds produce a gnatcatcher-like mewing call that descends in pitch when the bird is perched in the open. They also produce an elongated series of descending *tsee* notes, rapidly repeated, when chasing another bird. From low in the underbrush, they sing a cascade of clear notes, descending in pitch; one observer believes this is a courting call.

SIMILAR SPECIES: Berylline Hummingbirds have the same bright green gorget extending to the lower breast. Beryllines have a black maxilla; an iridescent violet tail; and a conspicuous cinnamon bar from the inner primaries across the secondary wing coverts, which can be seen when the bird is either perched or in flight. The Cinnamon Hummingbird might look very much the same from the back, but there is no green on its ventral side. The whole undersides are cinnamon washed.

DISTRIBUTION: Buff-bellied Hummingbirds are resident and breed in the coastal regions of the Gulf of Mexico, from Belize, Guatemala, and the Yucatán Peninsula north to Victoria, Texas, and west along the Rio Grande to Mission, Texas. In winter, the US population apparently increases and spreads along the Gulf Coast north and east through southern Louisiana, with scattered records farther northwest almost to Austin and the Texas Hill Country, and east into Florida. They prefer the semiopen habitat of broadleaf forests in riparian areas along rivers and creeks. They are also found in scattered groves of oak trees (*Quercus* sp.) and thorn scrub along the Lower Rio Grande Valley; in woodlands with grass between scattered hammocks of taller trees; and in parklands and residential areas with similar habitat structures. Near Victoria, Texas, they have a strong preference for the nectar of Turk's cap (*Malvaviscus arboreus* var. *drummondii*) and may appear in numbers when this plant is flowering. Other favored species include sweet acacia (*Acacia farnesiana*), locally known as huisache, and Texas wild olive (*Cordia boissieri*), locally known as anacahuita. Postbreeding wandering along the Texas coast and along the Gulf Coast east to Florida has increased in recent years.

MIGRATION: In the breeding areas from southern Texas to north of Corpus Christi, Texas, the numbers of birds begin to disperse starting in October. Most of the remaining birds concentrate within one hundred miles of the coast. In February, birds begin leaving wintering areas along the Gulf Coast and Mexico, moving inland to nest. Some remain north of Corpus Christi until mid-May, when Turk's cap blooms, and then they disperse to breed. Others from Victoria and Corpus Christi south may start nesting as early as late February and March. Birds departing from this area leave in late October, and by November, only the resident birds remain.

COURTSHIP AND NESTING: Males may display during courting, but it is not certain if the displays are to attract females or to scare off rival males. In areas south of Corpus Christi with adequate food, females will start nesting in February or early March, as soon as they arrive on breeding grounds. Farther up the coast, where the birds depend heavily on Turk's cap, they may not start nesting until they move inland in mid-May. They may have two broods: one early in March and a second beginning in May. Recently fledged young are seen from May through July. Some birds delay breeding until later in summer, or else they may renest later; juveniles have been seen as late as October. Nests are built by females in small trees and shrubs from two to seven meters above the ground. The cup-shaped nest is similar to that of other hummingbird species and is made of thin flexible plant material woven together and bound with spiderweb. The interior is padded with plant down, and the exterior is decorated with lichens, leaves, or bark for camouflage.

BEHAVIOR: Buff-bellied Hummingbirds are larger than the other two species of hummingbirds commonly found in their North American range in southern Texas, the Ruby-throated and the less common Black-chinned. Therefore, they dominate feeders and flower patches and tend to chase other birds away. They may aggressively defend a feeder or flower patch even when there are many birds in the area. However, when Ruby-throateds migrate through in large numbers, Buff-bellied Hummingbirds "give up" and apparently do not waste energy fighting off groups of birds. Their attitude, like that of many hummingbirds, appears to be solitary most of the year.

NUTRITION AND MOLT: Buff-bellied Hummingbirds are opportunistic nectar feeders, taking advantage of available nectar anywhere they can find it. They are not restricted to typical tube-shaped flowers, but instead feed anywhere there is a good supply of nectar. In southeastern Texas, they consume nectar from shrimp plant (*Justicia brandegeana*), coral bean (*Erythrina herbacea*), Turk's cap, aloe (*Aloe* sp.), and sage (*Salvia* sp.). They are easily attracted to artificial feeders. Like other hummingbirds, they glean insects from leaves, flowers, and branches of vegetation, in addition to fly-catching them. In winter, they are known to drink from sap wells opened by woodpeckers, in the few areas where sapsuckers and woodpeckers coexist with Buff-bellieds. Body and wing molt begins after breeding is finished for the season. We have no information on the sequence of molt of body, tail, or wing feathers.

VIOLET-CROWNED HUMMINGBIRD, VCHU (*Amazilia violiceps*) Colibrí Corona-violeta (Sp.) *Amazilia* = word of unknown origin; *violiceps* = violet-colored head

IDENTIFICATION: Because of its pure white underparts, red bill, and blue-violet crown, the Violet-crowned Hummingbird is easily distinguished from all other hummingbirds in North America. All ages and both sexes have similar plumages, making the species easy to recognize during any season.

ADULTS: Sexes cannot be differentiated in the field by any external character of which we are aware. Gravid females can be sexed in the hand. The bill is red at the base with a black tip. Iridescent bluish-purple feathers cover most of the crown and sometimes extend down around the eye, ear, and nape of the neck. The back is gray green or bronzy green, often appearing gray. The tail is bronze colored and usually without obvious white tips. Some adult females have white at the tip of r5.

JUVENILES: Young birds have more black on the bill than adults. The crown may lack any of the blue-violet feathers or bright iridescence of adults. Some tail feathers have white tips. The maxilla is corrugated for the first month or two after fledging. Back feathers have grayish fringes, which quickly wear away.

Violet-crowned Hummingbird

Adult Violet-crowned Hummingbird.

SIMILAR SPECIES: No other North American species is similar.

DISTRIBUTION: Violet-crowned Hummingbirds have a relatively small and well-defined range, from southeastern Arizona and southwestern New Mexico south through the sky islands into northern Mexico; down the Sierra Madre Occidental and interior of Mexico; and along the Pacific coast south to northern Oaxaca. Violet-crowneds breed only as far north as eastern Arivaipa Canyon, Arizona, but they are more common to the south along Sonoita Creek and streams bordered by sycamore trees (*Platanus* sp.), and in the sky islands of southeastern Arizona and southwestern New Mexico. They have been seen in southern California; near Amarillo, Texas; and in southwestern Texas. Their principal range is western Mexico. In Arizona, Violet-crowneds prefer riparian areas with Arizona sycamore (*Platanus wrightii*), in which most birds nest. In Mexico, they occur in drier habitats, thorn forests, and oak woodlands, and around city parks and gardens.

Medium-Sized Hummingbirds

MIGRATION: Violet-crowneds migrate north into Arizona and New Mexico with the flowering of century plants (*Agave* sp.), from which they gather nectar in June. Fall migration of the northern population occurs from September through early November. Some birds spend the winter in southeastern Arizona.

COURTSHIP AND NESTING: No elaborate courtship rituals or flights are known for this species. Sexes cannot be told apart in the field. Breeding begins in April. Nests are built mostly on sycamore branches from seven to twelve meters above the ground; they are made of downy material, which is stuck together with spiderweb and then decorated with lichens, weed seeds, and sometimes green leaves. Young appear in some areas as early as mid-April and are seen through mid-September. The two young are fledged about forty-two days after egg laying, with the specific time for incubation unknown.

NUTRITION AND MOLT: Violet-crowneds feed on both plant nectar and insects caught in the air and gleaned from foliage. In general, they feed higher in the canopy than most smaller hummingbirds, but they will come near the ground to feed on available nectar-producing flowers. Some plants the Violet-crowned prefers are ocotillo (*Fouquieria splendens*), bouvardia (*Bouvardia* sp.), sage (*Salvia* sp.), prickly pear (*Opuntia* sp.), and many species of century plant. It will also attend artificial feeders. While it will guard patches of nectar plants and assert its dominance over smaller species, it is not an aggressive species. There is much yet to learn about this species' ecology and natural history. Some adults molt primary wing feathers in summer, but we do not have good information on the rest of the molt, which must occur from late fall to early spring.

Violet-crowned Hummingbird

Adult, dorsal view.

Adult, ventral view.

Juvenile, dorsal view.

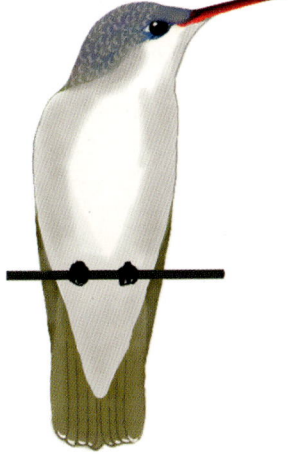

Juvenile, ventral view.

Medium-Sized Hummingbirds

The tail of an adult is uniform olive green to olive brown. There are very thin gray-white tips to each tail feather r2 through r5.

The tail of a hatching year (juvenile) is also olive green or brown, but it has definite white tips to at least r2 through r5.

Violet-crowned Hummingbird

Adult female Violet-crowned Hummingbird. We know this is a female because it was gravid. (It had an egg visible in the belly.)

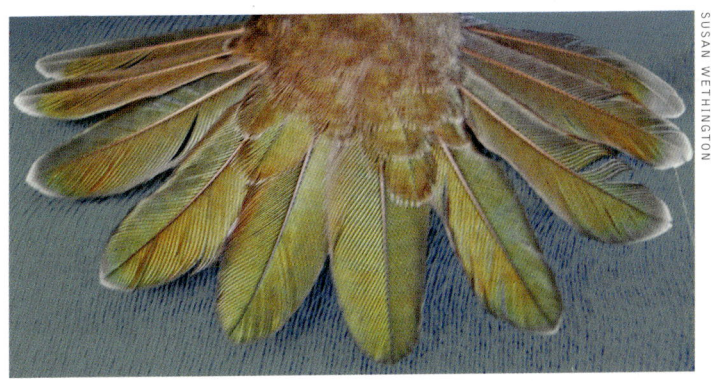

This adult Violet-crowned has white tips on one r5, demonstrating that white tail tips are not a reliable character to differentiate ages of the species.

Medium-Sized Hummingbirds

This adult has an all-red bill with a less well-defined black tip.

This adult has an all-red bill with a well-defined black tip, a blue to violet crown depending on the light, and all-white underparts.

Violet-crowned Hummingbird

Adult in the field. Sometimes the crown is very purple or lilac; from a different light angle, it is blue.

This juvenile has a less well-defined black bill tip and only a sprinkling of blue feathers on the head. Otherwise it resembles the adult. Pale fringes to each feather can be seen on the nape—further evidence of the bird's age.

Medium-Sized Hummingbirds

A bird faces a direction that does not show its blue-violet crown, but it is still recognizable by its all-olive brown back, bright white underparts, and red bill with black tip.

Adult Violet-crowned. The bird is probably an adult, but it still shows a white tip to the outer tail feather.

Violet-crowned Hummingbird

Juvenile Violet-crowned, showing the lack of blue in the crown and pale fringes to the crown and back feathers.

This maneuvering adult shows off its bronze-green tail with light gray (but not pure white) tips. Note the ability of the bird to make extreme maneuvers in flight, by bending its wing and adjusting its flight feathers to make the turn.

Medium-Sized Hummingbirds

In an attempt at camouflage, the nest of a Violet-crowned Hummingbird is covered with lichens and bits of green leaves.

SMALL HUMMINGBIRDS

Birds with a total length of about 3½ inches or less, including the bill

WHITE-EARED HUMMINGBIRD, WEHU (*Hylocharis leucotis*) Colibrí orejiblanco (Sp.) *Hylocharis* = wood beauty (Boie); *leucotis* = white-eared (Viellot)

IDENTIFICATION: White-eared Hummingbirds appear stubby, with short necks and large heads. The short, straight bill is red with a dark tip in adult males. In adult females, the maxilla is black with some red at the base, while the mandible is reddish with a black tip.

ADULT MALE: The bold white stripe from above the eye back along the head contrasts with the dark blue-green head (which often appears black). The throat is heavily spotted with dark green spangles, which extend—along with green feathers with white margins—to the upper breast and along the flanks. The back is dark green, blending into cinnamon or rust toward the tail. The tail is square at the tip. There may be some white on the tips of the outer two or three rectrices (r3, r4, and r5).

Adult male White-eared Hummingbird.

White-eared Hummingbird

Adult male, dorsal view.

Adult male, ventral view.

Adult female, dorsal view.

Adult female, ventral view.

Small Hummingbirds

Juvenile male, dorsal view.

Juvenile male, ventral view.

Juvenile female, dorsal view.

Juvenile female, ventral view.

White-eared Hummingbird

The tail of an adult male is all dark with olive-green central feathers (r1) and darker, almost black, r2 through r5. At least r2 and r3 show a bronze-green iridescence in sunlight. All tail feathers are relatively short and wide.

The tail of an adult female also has bronze-green central feathers and darker blue and green r2 through r5. The green on the lateral margins of at least r2 and the basal portions of r3 and r4 shows in good light. There is a very small white tip on r2 (sometimes there is none) and white tips increasing in size on r3, r4, and r5. In both sexes, you can see the rufous fringes of the upper tail coverts. The juvenile tail of both sexes resembles the tail of the adult female.

Small Hummingbirds

Adult male.

The adult male White-eared Hummingbird is easy to identify, with its bright red bill with a black tip and its dark head with a brilliant white stripe from above and behind the eye, down past the ear to the neck. Only when the sun is shining on the bird's head and the bird is facing the sun do you see its dark blue feathers. Usually only the blue green of the gorget is seen, as in these photos.

White-eared Hummingbird

Small Hummingbirds

Adult females have much-reduced green spangles in the gorget, and the crown is lighter and browner than that of the male.

Adult female, showing the green back, brown crown, and conspicuous postocular white stripe.

White-eared Hummingbird

Juveniles resemble adult females. They have cinnamon fringes on the crown and back feathers for a while after fledging, as can be seen on the head of this bird.

ADULT FEMALE: The female has a dark face patch (mask), which is not as dark as the male's. It is in sharp contrast to the white supercilium. The crown is browner than and not as dark as that of the male. The gorget, upper breast, and flanks have few to many green feathers. The back is dark green. The tail is almost square, with bronzy-green central feathers (r1) and darker, white-tipped outer feathers (r2–r5).

JUVENILES: Resemble adult females. Corrugations on the maxilla last a month or more. The crown and back feathers are tipped with buff for several weeks after fledging.

JUVENILE MALE: The crown is paler than that of the adult male or female. There are usually some green spangles on the throat. The outer two or three pairs of rectrices have pale tips.

JUVENILE FEMALE: Resembles the adult female. The crown is paler brown than that of the adult female. The white supercilium is not as bright as the adult female's. There are fewer throat and breast spots, and the tail tips have more white.

SIMILAR SPECIES: The Broad-billed Hummingbird female is a longer and thinner bird, with a smaller head and a longer bill. Its white eye stripe and black mask are much narrower than the White-eared's. The Broad-billed's tail is larger and bluer, with large white spots on r5. Xantus's Hummingbirds have cinnamon underparts and tails.

DISTRIBUTION: The White-eared Hummingbird is a Mexican species that comes into the United States regularly only in southeastern Arizona and the Davis Mountains of Texas. It has been seen in southern New Mexico. Most of the population occurs in higher elevations in the Sierra Madre ranges of Mexico and the humid forests of southern Mexico, south to Nicaragua. In Arizona, it is almost annual. It breeds occasionally in the sky islands of the Chiricahua, Huachuca, and Santa Rita Mountains. White-eared Hummingbirds prefer juniper (*Juniperus* sp.), pine/oak (*Pinus/Quercus* sp.), and evergreen oak forests at higher elevations in both Mexico and Arizona, with clearings that support nectar-producing flowers.

MIGRATION: In Mexico, migration is primarily altitudinal. The small Arizona population may arrive as early as March, but it usually does not arrive until late April. It departs starting in August, although some birds remain until October.

COURTSHIP AND NESTING: Males return first and set up feeding territories. As the females arrive, males gather in groups and display and call, similar to lek activity in grouse. The female approaches a male and they fly to a nesting site, where they copulate. The male departs, and the female builds the nest and raises the young. In central Mexico, White-eareds nest from March through August. Two eggs are laid, and incubation lasts fourteen to sixteen days. Young are fledged after twenty-three to twenty-six days.

NUTRITION AND MOLT: The White-eared usually feeds close to the ground. In the pine forests of Mexico, it is the hummingbird most often seen at eye level. In Arizona, it comes to feeders, which may be one of the principal reasons it has flourished there. White-eareds feed on nectar-producing flowers and small insects caught in the air or gleaned from vegetation. Little is known about the timing of molt in Arizona birds.

BROAD-BILLED HUMMINGBIRD, BBLH (*Cynanthus latirostris*) Colibrí piquiancho (Sp.) *Cynanthus* = dog flower (Swainson); *latirostris* = wide-billed (Swainson)

IDENTIFICATION: Broad-billeds are small blue-green hummingbirds. They usually appear stretched out, with long necks, sleek bodies, and red bills.

Adult male Broad-billed Hummingbird.

Small Hummingbirds

ADULT MALE: The back is blue green, the gorget and upper breast are blue, the belly is green, and the undertail coverts are white. The long, slightly decurved red bill has a black tip. The tail is dark blue and notched.

ADULT FEMALE: The crown and back are green. The maxilla is black, and the mandible is pinkish to red at its base. There is a white supercilium and a dark mask back of the eye. The large, dark blue notched tail has large pale to white spots on the tips of the two outer feathers (r4 and r5).

JUVENILES: Resemble adult females. The grooves on the maxilla and the pale cinnamon margins to the crown and back feathers remain for some weeks after hatching.

JUVENILE MALE: There may be some blue spangles on the throat. The base of the bill is red, and the dark blue tail has very thin white or pale gray tips. As the summer progresses, an increasing number of blue spangles appear in the gorget, and the belly darkens with blue-green feathers.

JUVENILE FEMALE: There is less red on the bill than in the juvenile male. Once the grooves and buff back have disappeared in late summer, it is not possible to distinguish adult from juvenile females.

VOICE: Their common call note is a chattered *je-jit*. It is most often given singly, but it can be joined to make a longer chatter call.

SIMILAR SPECIES: Except for the larger Accidentals Green Violet-ear and Green-breasted Mango, which are dark green to blue green, there are no male hummingbirds similar to the Broad-billed male. Females are often mistaken for White-eared Hummingbirds, but the White-eared is heavier, with a larger head, a much shorter bill, and more spotting on the upper breast and flanks. The White-eared's white eye stripe is wider, as is the black

Broad-billed Hummingbird

Adult male, dorsal view.

Adult male, ventral view.

Adult female, dorsal view.

Adult female, ventral view.

Small Hummingbirds

Juvenile male, dorsal view.

Juvenile male, ventral view.

Juvenile female, dorsal view.

Juvenile female, ventral view.

Broad-billed Hummingbird

Adult male tail. Note the all-dark blue notched tail. The central feathers are shorter than the outer feathers.

Adult female tail. Note the dark tail with white on the tips of r4 and r5. The white on the tip of r3 has been worn away during nesting. Sometimes there is some white at the tip of r2.

Small Hummingbirds

Juvenile male tail. Note the dark blue tail, which is the same as the adult male tail except that there are white tips on r3 through r5.

Juvenile female tail. Note the dark tail, which is almost identical to that of the adult female, with white on the tips of r3 through r5.

Broad-billed Hummingbird

Note the adult male's red bill with black tip, violet-blue gorget and neck, blue breast, green belly and back, and dark tail. The only other hummingbirds that are predominantly blue green are the accidentals Green Violet-ear and Green-breasted Mango, both of which are much larger and have black bills.

Small Hummingbirds

This adult female has a white streak back of the eye, which is not as wide, long, or bright as the postocular streak of a White-eared Hummingbird. The bill of this bird is longer, and the throat lacks dark spots. The back and upper tail coverts are more blue green, not olive green, with rusty fringes like in the female White-eared.

Broad-billed Hummingbird

The red on the bill of this adult female is hard to see, but it is visible at the base of the mandible. Note the lack of color on the undersides, the dusky mask, and the light line over and behind the eye. Also note the long and thin body shape. Compare this bird to a female White-eared Hummingbird, which is much chunkier, with a shorter bill, darker mask and face, and brighter and more extensive white postocular stripe. This female is in heavy body molt, although it has finished molting its flight feathers.

In this photo of an adult female taken in September, the red on the mandible is clearly visible. The bird is still molting its primary flight feathers.

Small Hummingbirds

Juvenile male, showing the slowly developing blue and green gorget and neck feathers.

Another juvenile male, showing its developing red mandible and gorget.

Broad-billed Hummingbird

A Broad-billed Hummingbird nest, suspended from a small tree in Madera Canyon in Arizona. Nests are often in riparian areas and are constructed to resemble debris left in the brush after annual monsoon floods.

Note the scaly appearance to the head and back feathers of this juvenile female. Red is only visible on the underside of the bill.

Small Hummingbirds

Adult male in flight. Its bright green belly, blue throat, and red bill make identification of this bird relatively easy.

auricular area. Its tail is smaller, more squared at the tip, and not as blue as the Broad-billed female's tail.

DISTRIBUTION: The Broad-billed Hummingbird regularly occurs in the United States only in southeastern Arizona and the extreme southwest of New Mexico. Its principal breeding range is from southeastern Arizona along Mexico's Sierra Madre Occidental south to Jalisco. There are several records of breeding in the Chisos Mountains of Big Bend National Park, Texas. Most of the southern populations in Mexico are resident; those in the northern extent of the breeding range migrate annually into Mexico. A few birds overwinter in Arizona. In Arizona, most breed at lower elevations in riparian areas dominated by sycamore (*Platanus* sp.) and cottonwood (*Populus* sp.) trees. Some breed at middle elevations in the sky islands, along canyons that support the same trees. In Mexico, they breed from low to high elevations, from riparian areas through thorn scrub, up to higher-elevation oak (*Quercus* sp.) and pine (*Pinus* sp.) forests.

MIGRATION: Males come north into Arizona in early to mid-March. They are soon joined by the females, with most of the population settled by late April. Adult males begin to depart in August, followed by the females and juveniles in early to mid-September. Except for a few that overwinter, birds are usually gone by the first of October.

COURTSHIP AND NESTING: Breeding begins soon after the birds arrive, with egg laying occurring from mid-March into April. Second broods are common, depending on food supply. Nests are usually built close to the ground, on branches that hang down below the shrub canopy. The loosely constructed nest resembles flood debris. Only the females feed the nestlings, which are fledged in late March (rarely) through August.

NUTRITION AND MOLT: Like other small hummingbirds, Broad-billeds depend largely on nectar from a wide variety of local plants. They also catch insects from the air and pick them and spiders from leaves and flowers. One study of their daily activity shows that 36 percent of daylight hours are spent perching, 54 percent feeding, and 10 percent in flight. Males start molting body and flight feathers in July, and females follow after the young are fledged. Alternate and basic plumages are the same for adults.

LUCIFER HUMMINGBIRD, LUHU (*Calothorax lucifer*)
Tijereta Norteña (Sp.) *Calothorax* = beautiful breast (but the author meant gorget) (Gray); *lucifer* = light bearer or torchbearer (Swainson)

Small Hummingbirds

IDENTIFICATION: This species is fairly easy to identify by its long, decurved bill; relatively large head; and narrow, rather short, forked tail. Often the tail is compressed into a point. It is a xerophile, preferring the dry desert arroyos and canyons—frequently those with riparian growth of oak (*Quercus* sp.), cholla (*Opuntia* sp.), ocotillo (*Fouquieria splendens*), and century plant (*Agave* sp.).

ADULT MALE: The gorget is brilliant rose purple, but the color may range from pinky violet to purple to black depending on the light angle. A white postocular stripe separates the gorget from the crown. The undersides are light with green and buffy flanks. The dark tail is deeply forked.

ADULT FEMALE: The crown and back are green. There is a dusky mask back of the eye and a buffy postocular stripe. The throat is very pale, the breast and belly are buffy, and the flanks are rusty. The tail is notched. When spread, the central tail feathers (r1) are short and bronzy green. The outer three pairs of tail feathers are longer; each has a variable amount of rufous at the base, followed by black, and ending in a white tip.

Adult male.

Lucifer Hummingbird

Adult male Lucifer Hummingbird, dorsal view.

Adult male, ventral view.

Adult female, dorsal view.

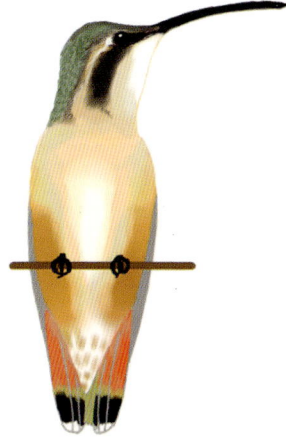

Adult female, ventral view.

Small Hummingbirds

Juvenile male, dorsal view.

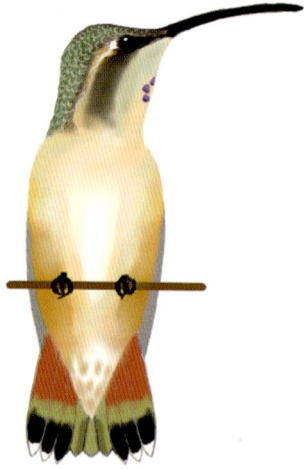
Juvenile male, ventral view.

Juvenile female, dorsal view.

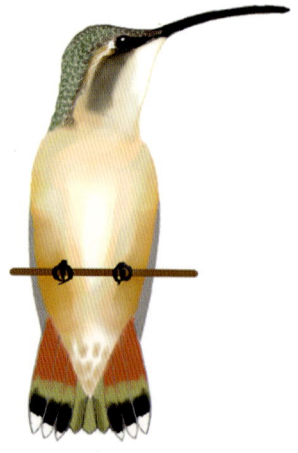
Juvenile female, ventral view.

Lucifer Hummingbird

Spread tail of an adult male. All feathers are dark with some brown trim. The longest is r4 and the shortest is r1, making the tail deeply forked.

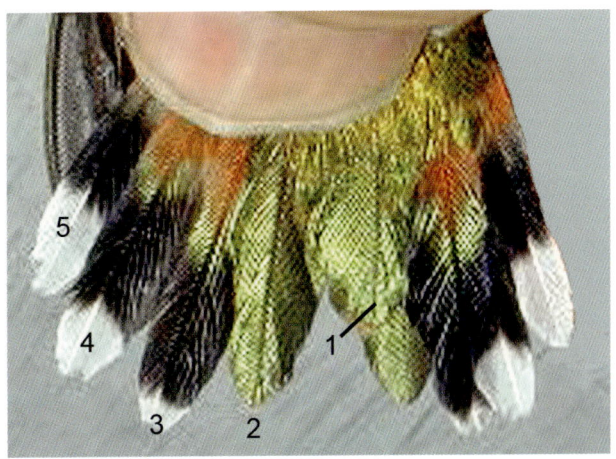

Juvenile male tail, showing the large white tips to r3 through r5 and the all-green r1 and r2. Again, r1 is very short and hard to see against r2 below it.

Small Hummingbirds

Adult female tail (right half). Note the rusty bases to r3 through r5, all-green r1 and r2, and very short r1 (over r2 in this photo), the latter giving the tail a forked shape when partly opened.

Adult male. Note the long, decurved bill on a "large-headed," slim body, white around the magenta/violet gorget, dusky sides of the breast, and cinnamon flank.

Lucifer Hummingbird

Juvenile male. Note the violet spangles beginning to appear in the gorget and the scaly appearance of juvenile feathers on the crown.

Adult female. In addition to the downcurved bill, note the pale cinnamon wash to the face and throat; black mask; and rusty flanks, belly, and breast.

Another adult female, showing the buffy supercilium and postocular stripe.

Small Hummingbirds

Adult male in flight. The long downcurved bill and purple gorget make identification of this bird relatively easy. Note that the compressed tail is very narrow.

JUVENILES: Resemble adult females. The corrugations on the maxilla and light fringe tips of the crown and back feathers will disappear after a few weeks. The buffy undersides, rusty flanks, and tail pattern are the same as those of the adult female.

JUVENILE MALE: Young males may have some to many purple spangles on the throat.

JUVENILE FEMALE: Young females have more white at the tips of tail feathers r3 through r5 than adult females. Once the bill grooves and buff back are gone, distinguishing adult from juvenile females may be impossible.

SIMILAR SPECIES: Bumblebee Hummingbird adult males have similarly colored gorgets, but they are much smaller and have short, straight bills. The only other species with a markedly decurved bill is the much larger Green-breasted Mango.

DISTRIBUTION: While the Lucifer Hummingbird is a bird of south-central Mexico, several populations exist in North America during the breeding season. The largest population is in the Chisos Mountains of Big Bend National Park, Texas. Other smaller populations are in the Christmas and Davis Mountains, Texas; southwestern New Mexico; and, regularly, southeastern Arizona.

MIGRATION: Migration begins when males, and then females, come north in March and April. Most birds depart their northern range by the end of September.

COURTSHIP AND NESTING: Males may defend both a flower patch and a nesting area. Unlike other hummingbirds, the male will court the female as she is building the nest. Displays consist of a shuttle in front of the female, usually within touching distance of the female's bill. Then the male begins an upward flight in wide S-turns of up to twenty to thirty meters. He proceeds to dive vertically to within one to two meters above the ground (where nests are usually located). Each display is accompanied by easily heard sounds created by the wings and tail. Nesting begins in the North American population in April and may continue through July. Young may appear as early as late April, but more commonly they are not seen until July. Nests are built on branches of oak, ocotillo, or other desert plants. It is not uncommon to have nests located within a few hundred meters of each other. Two eggs are laid, the second two days after the first. Incubation by the female lasts about fifteen days, and the young fledge after twenty-two to twenty-four days. The female continues to feed the young for as many as nineteen days after leaving the nest.

NUTRITION AND MOLT: More is known about the main population in Mexico; only a few studies have been conducted on the North American population. The primary foods are nectar and small insects and spiders. Birds seek out any flowering plants in their dry environment, including ocotillo, century plant, pentsemon (*Penstemon* sp.), hummingbird bush (*Anisacanthus quadrifidus*), paintbrush (*Castilleja* sp.), desert willow (*Chilopsis linearis*), cholla, and yucca (*Yucca* sp.). They fly out from an exposed perch to catch small flying insects, and they are not as likely to glean insects from leaves or branches as other hummingbirds.

RUBY-THROATED HUMMINGBIRD, RTHU (*Archilochus colubris*) Colibrí Gorjirrubí (Sp.) *Archilochu*s = first among the birds (Reichenbach); *colubris* = serpent (probably a misspelling of *colibri*, the French word for hummingbird)

IDENTIFICATION: The Ruby-throated is the only common hummingbird in eastern North America. It has been documented only a few times in the southwestern states west of Texas. Both Ruby-throated and Black-chinned Hummingbirds have the following wing feature: the anterior and posterior vanes of primaries 7 through 10 are about equal in width, while the anterior vanes of p1 through p6 are narrower than the posterior vanes. The width of p10 at 5.0 mm from the tip is less than 2.5 mm.

Adult male.

Ruby-throated Hummingbird

Adult male Ruby-throated Hummingbird, dorsal view.

Adult male, ventral view.

Adult female, dorsal view.

Adult female, ventral view.

Small Hummingbirds

Juvenile male, dorsal view.

Juvenile male, ventral view.

Juvenile female, dorsal view.

Juvenile female, ventral view.

Ruby-throated Hummingbird

Tail of an adult male. Note that the central feathers (r1 and r2) are bronze green, and r3 through r5 are black and longer than r1. The longest feather is usually r5, which is also relatively thin.

Tail of an adult female. Note that only r1 is all green, while the other feathers are black and have green only at the base. R3 through r5 have large white tips. The tip of r5 is dull pointed, not rounded.

Small Hummingbirds

The left half of the tail of a juvenile male Ruby-throated. It closely resembles the tail of the adult female. Juvenile female tails have rounded, not dull pointed, r5 tips.

Wing of an adult male. Note the notch at the tip of the posterior vane of p5. The same pattern is also on p6 and p4. No other hummingbird in North America has this character.

Ruby-throated Hummingbird

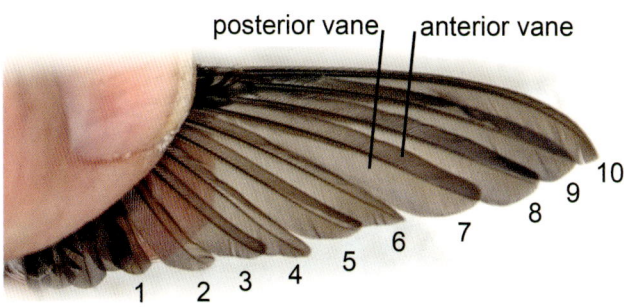

Wing of an adult female. Note that p6 is sharp pointed and not notched, as is the male's feather. P4 and p5 are somewhat flat across the tip, in contrast to p2 or p7.

When the adult male Ruby-throated is perched, its tail forms a deep fork, as the central tail feathers are much shorter than the outer ones.

Gorget and breast of two adult females, showing the lack of definite spotting. Most birds are white like the bird at left, but some are duller; they rarely have one or more spangles.

Small Hummingbirds

Adult male. With its head turned sideways to the light, the gorget appears almost black.

Adult male, showing its forked tail, dark red (in this light) gorget, white breast, and white vent band.

Adult female, showing its clear-white gorget, breast, and belly. Compare the females in these images to female Black-chinned Hummingbirds.

Ruby-throated Hummingbird

ADULT MALE: The crown and back are green. The face is blackish. The brilliant ruby-red gorget is bordered by black at the chin. There is a distinct white postocular spot. The white of the breast extends to the shoulder, making a distinct contrast among the gorget, nape, and back. The flanks and lower belly are greenish gray. The undertail coverts are white to buffy white. The blackish tail is deeply notched, and the outer tail feathers are long and pointed. Primaries 6 and 5 have a dent or notch in the tip of the posterior vane.

ADULT FEMALE: Females are larger than males. The crown and back are green. The entire undersides are very white, with little if any spotting. The face is dusky and has a small white postocular spot. The flanks are gray green, sometimes with light buff. The tail is notched; the central pair of feathers (r1) is solid gray green. The second (r2) to fifth (r5) rectrices are gray green at the base, with a blackish band toward the tip. R3 through r5 have white tips. R5 ends in a dull point. The tip of the posterior vane of p5—and, more so, that of p6—is cut at an angle. P5 and p6 are not as rounded as the other primaries.

JUVENILES: Resemble adult females. The maxilla has fine corrugations lasting for a month or more after fledging. The back and crown feathers have pale fringes to their tips, which wear away by late summer.

JUVENILE MALE: There are some dark feathers on the throat; occasionally there is a red spangle. The notched tail resembles that of the adult female, but r5 is pointed.

JUVENILE FEMALE: Larger than the juvenile male, with a longer bill. The notched tail is like the adult female tail, except that the tip of r5 is rounded, not pointed. There may be a slight white tip on r2.

SIMILAR SPECIES: Females are easily confused with Black-chinned females. Ruby-throated females are greener above and whiter on the gorget and below. The tips of p5 and p6 on the

Ruby-throated are "cut off" instead of rounded. P10 is narrower than the p10 of a female Black-chinned Hummingbird.

DISTRIBUTION: The Ruby-throated Hummingbird is a bird of the eastern United States. It rarely comes west of the central plains. In recent years, it has been seen more often in the western United States, although there are no records of breeding there yet. It has been appearing more frequently in western Texas, but not as often in either New Mexico or Arizona. In Canada, however, it breeds across southern Manitoba into southern and central Saskatchewan. The Ruby-throated winters at the southern tip of Florida, on the Yucatán Peninsula, in southern Mexico and Guatemala, and along the Pacific coast of Mexico, from Nayarit south.

MIGRATION: Ruby-throateds may move south in fall along the Texas coast and down through Mexico, or they may cross the Gulf of Mexico from Florida or any of the Gulf states to Mexico. They return the same way in spring. The seeming impossibility of this migration across many miles of open water has led to various myths, according to which hummingbirds are carried on the wings or backs of larger birds, such as geese or swans. Because the breeding distribution extends thousands of miles from south to north, the timing of events differs greatly. Arrival of birds from the south may occur in early March, but in the more northern areas of the breeding range, arrival might not take place until late May. Similarly, southward departure dates are earlier in the northern than in the southern portions of the breeding range.

COURTSHIP AND NESTING: Nesting follows the same latitude-dependent schedule as does migration. Ruby-throated males are very aggressive, and they dive on and pursue any intruder into their feeding territory. They begin courtship with a similarly agonistic display, consisting of a U-shaped looping dive that starts twelve to fifteen meters above the ground. If the female is perched, the male will shuttle back and forth in front of her, keeping the gorget feathers elevated and facing the sun to give maximum reflective brightness. Nests are built by the female, usually near the end of a downward-sloping branch of oak (*Quercus* sp.), hornbeam (*Carpinus* sp.), birch (*Betula* sp.), poplar (*Populus* sp.), hackberry (*Celtis* sp.), or pine (*Pinus* sp.), from one-half to fifteen meters above the ground. Nests are made of plant down, which is woven together with spiderwebs and decorated with lichens. Normally two eggs are laid, the second from one to three days after the first, with incubation beginning when the first egg is laid. This results in asynchronous hatching; one young is larger and more mature than the other. Incubation is twelve to fourteen days, and fledging occurs eighteen to twenty-two days after hatching.

NUTRITION AND MOLT: In the eastern United States, Ruby-throateds are the common hummingbirds at feeders. They arrive at their home feeder often within days of the date on which they arrived the prior year, and they also depart on the same schedule. Ruby-throateds feed on nectar-producing trees; flowering plants; sap wells; and insects caught in the air or gleaned from foliage, bark, and spiderwebs (including the spider). Adults may start molting body feathers by mid-May, but the summer molt is incomplete. They molt all body and flight feathers in late winter and early spring.

BLACK-CHINNED HUMMINGBIRD, BCHU (*Archilochus alexandri*) Colibrí Barbinegro (Sp.) *Archilochus* = first among the birds, name of a Greek soldier-poet (Reichenbach); *alexandri* = Dr. M. M. Alexandre (Bourcier & Mulsant)

IDENTIFICATION: The Black-chinned Hummingbird pumps or bobs its tail when hovering at flowers or feeders. It is the most abundant hummingbird in the western United States, rivaling the abundance of its eastern congener, the Ruby-throated Hummingbird. Male Black-chinned Hummingbirds are smaller than females, and in poor light, they are hard to distinguish from male Ruby-throateds at feeders. Females are very difficult to distinguish

Small Hummingbirds

Adult male Black-chinned Hummingbird.

from congeneric Ruby-throated Hummingbird females. The width of p10 at 5.0 mm from the tip is more than 3.0 mm.

ADULT MALE: The head appears almost all black, with a small white postocular spot. The gorget is black, but in good light, a violet crescent is visible at the lower margin of the gorget. The breast is dull white, and the flanks are grayish green. The all-dark tail is notched. The outer three pairs of feathers (r3–r5) are sharply pointed.

ADULT FEMALE: Females are noticeably larger than males. The back is grayish green, and the crown is often very gray or gray brown. The whole head appears very drab compared to that of a female Ruby-throated. The undersides are clear to slightly spotted. Often there are gray green spots on the throat. The flanks have some cinnamon below and gray green above. The tail is rounded, with white tips on the outer three pairs of feathers (r3–r5). The tip of the outer feather (r5) is dull pointed.

Black-chinned Hummingbird

Adult male Black-chinned Hummingbird, dorsal view.

Adult male, ventral view.

Adult female, dorsal view.

Adult female, ventral view.

Small Hummingbirds

Juvenile male, dorsal view.

Juvenile male, ventral view.

Juvenile female, dorsal view.

Juvenile female, ventral view.

Black-chinned Hummingbird

Adult male tail. Note the sharp points of r2 through r5. All of these feathers are black, with no white tips.

Adult female tail. Note that r3 through r5 have white tips, and the tip of r5 is pointed. Sometimes r2 has some white at the tip, but the white of r2 to r4 is often worn down after nesting, leaving r5 with the most white.

Small Hummingbirds

This single feather is an r5 from another adult female. It has a nice sharp point.

Juvenile male tail. Note that the tail is clean and dark, with little wear at the feather tips. From the scaly (buff) back and grooves on the culmen, you know that this bird is a juvenile. The pointed tip of r5 will tell you that it is a juvenile male.

Black-chinned Hummingbird

Juvenile female tail. Note that the tip of r5 is rounded, not pointed (unlike in the juvenile male's tail).

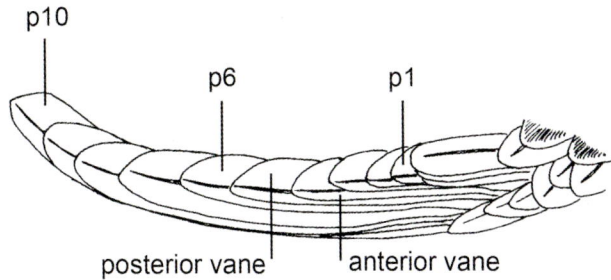

Wing of a female Black-chinned Hummingbird. Both Black-chinned and Ruby-throated Hummingbirds belong to the genus *Archilochus*, and their basic wing characteristics are the same. Note that the anterior and posterior vanes of primaries 10 to 7 are about equal in size. From p6 to p1, the anterior vane is very narrow compared to the posterior vane. In Costa's and Anna's Hummingbirds (genus *Calypte*), the anterior and posterior vanes remain about equal in width throughout all of the primaries. This can help to differentiate the female Black-chinned from the female Costa's Hummingbird, even in the field when the bird is perched.

Small Hummingbirds

Wings of Black-chinned Hummingbirds. Both show the *Archilochus* pattern of narrow anterior vanes and wide posterior vanes, especially on the inner primaries. Neither the male (*top*) nor female (*bottom*) shows the variations in tip shape of r4 through r6 seen in Ruby-throated Hummingbirds.

Black-chinned Hummingbird

The forked shape of a male Black-chinned's closed tail is similar to that of its congener, the Ruby-throated.

Note the black chin and violet crescent at the base of the gorget of this adult male. The bill is long and straight. The top of the head is black, as is the tail. The flanks are grayish to brownish green. This typical primary shape of the open wing shows the narrow anterior and wide posterior vanes to each of the inner primaries.

Small Hummingbirds

The violet of the gorget of this juvenile male is beginning to develop. The bird is browner—especially on the crown—than the adult, and except for the violet spangles, it resembles the juvenile female. The outer rectrix is more pointed, not rounded as it would be on a juvenile female.

Adult female Black-chinned Hummingbirds. The undersides are not as white as those of the female Ruby-throated, and there are more spots on the throat.

Black-chinned Hummingbird

Note the buffy margins to the crown and back feathers of this juvenile female, the brown cast to the plumage, and the "dirtier" gorget.

Adult female Black-chinned Hummingbird. Note the lightly speckled throat, unlike the cleaner Ruby-throated Hummingbird. Also note the green back and long, almost straight bill. The face is relatively "dirty" when compared to that of an adult female Ruby-throated. The primary feathers show the typical *Archilochus* shape.

Small Hummingbirds

Female Black-chinned Hummingbird on her nest.

Adult male in flight. If you can see both a purple band below the black gorget and a long black bill, you are safe in identifying this bird.

JUVENILES: Resemble a paler, grayer version of the adult female. Grooves on the maxilla and pale margins to the crown and back feathers disappear a month or more after hatching.

JUVENILE MALE: The throat is often heavily streaked. Sometimes there are a few black and/or violet feathers on the throat. If you can determine that the bird is a juvenile by grooves in the culmen and/or buffy edges to the crown or back feathers, check the shape of the tip of r5, which is dull pointed; the amount of white in r3 (none or only a slight bit at the tip); and the shape of the inner web of p1 through p6, which is emarginated.

JUVENILE FEMALE: The juvenile female is similar to the adult female after it loses the culmen grooves and scaly back feathers, except that the white tip of r5 is rounded, not pointed, on the juvenile female, and the shape of the inner web of p1 through p6 is rounded and not emarginated, as in the adult female. If the bird has no characteristics of a juvenile and the tip of r5 is pointed, the bird is probably an adult female.

SIMILAR SPECIES: In comparison to the female Costa's Hummingbird, the Black-chinned female is longer and thinner, with a much longer and slightly downcurved bill, a grayer crown (rather than the greener crown of Costa's), and a longer tail. It is usually also duskier below, and the wing pattern is different. Ruby-throated Hummingbird females have a narrower p10, and they are greener above and whiter below, without as much spotting on the throat. Black-chinneds do not have the Ruby-throated female's cut-off tips of primaries 5 and 6.

DISTRIBUTION: The Black-chinned Hummingbird is the commonest breeding hummingbird at middle elevations in the western states and provinces. It breeds from southwestern British Columbia and northwestern Montana south through central Idaho, western Colorado, New Mexico, and south-central and southwestern Texas, and through Arizona and interior California south to northern Baja

Small Hummingbirds

California, Sonora, and northwestern Chihuahua. Black-chinneds winter in northern, central, and west coastal Mexico, and a few usually remain in southern Texas in winter. North of Mexico, they prefer a mixed habitat of open grassland or desert and trees, as well as closed deciduous forests usually near streams or riverbeds.

MIGRATION: Males start their northward return in March, followed by females in April. They may not reach their northernmost breeding areas until May. Adults start moving south in early August, with peak southward movement occurring in mid-August. Most birds, including juveniles, have left for central Mexico by early October.

COURTSHIP AND NESTING: Courting males use a dive display in front of a female, which produces a rapid series of bell-like notes at the bottom of the dive. They also use a shuttle display. In the south, usually two or more nesting attempts are made each season. Females construct a cup-shaped nest measuring about 1½ inches across and ¾ of an inch deep, in hackberry (*Celtis* sp.), juniper (*Juniperus* sp.), evergreen oak (*Quercus* sp.), or cottonwood (*Populus* sp.) trees, and sometimes in sycamore (*Platanus* sp.) or willow (*Salix* sp.). They usually build the nest between six and twenty feet off the ground, building higher when larger trees are present. Often the nest is placed under a protective leaf, on the fork of a drooping branch. The nest is made of plant down, which is held together with spiderweb silk and often with pieces of lichen added to the outside of the cup. The female lays two eggs, typically one to two days apart; the first egg is often laid before the nest is complete. Incubation—by the female only—is from twelve to fourteen days, when the unfeathered young hatch. Young remain in the nest, fed by the female alone, for about twenty-one days until fledged. Fledged young appear after the first of June. Both adult and young birds wander away from the nest site before southward migration.

NUTRITION AND MOLT: Black-chinneds feed on nectar-producing flowers, e.g., verbena (*Lantana* sp.), century plant (*Agave* sp.), ocotillo (*Fouquieria splendens*), and tree tobacco (*Nicotiana glauca*), as well as on insects and arachnids from the ground to high in the trees. Often they fly-catch insects in midair. They are frequent visitors to nectar feeders, and they appear to be less wary of human presence than other hummingbirds. While some individuals commence body and primary molt on the breeding grounds, almost all will complete a full body and flight feather molt in winter in Mexico.

ANNA'S HUMMINGBIRD, ANHU (*Calypte anna*) Colibrí de Anna (Sp.) *Calypte* = veiled, referring to the combined red crown and gorget (Gould); *anna* = nineteenth-century Italian Duchess of Rivoli Anna de Belle Masséna (Lesson)

IDENTIFICATION: Anna's Hummingbirds are heavy-bodied hummingbirds, weighing about three-quarters of a gram more than Ruby-throateds or Black-chinneds. They appear chunky when perched, with shorter bills than *Archilochus* hummingbirds. The anterior and posterior vanes of the inner primaries are about the same width, unlike those of Black-chinned or Ruby-throated Hummingbirds.

ADULT MALE: The nape, back, and upper tail coverts are dark golden green. The crown and gorget flash electric rose pink to violet red to gold in sunlight. When the bird is not facing you, its head may appear black. There is a conspicuous white postocular spot. Sometimes the gorget has "ears," long individual spangles that project outward and backward along the neck. The bill is relatively short and straight. The undersides are heavily marked with green and gray—especially along the flanks—giving them a "dirty" look. The tail is deeply notched, with narrow outer rectrices and rounded feather tips.

ADULT FEMALE: There are a few to many red spangles in the center of the throat. The crown and back are dull green. The flanks are heavily spotted with gray and green, again making them

Small Hummingbirds

Adult male Anna's Hummingbird. Note the full rosy pink-red gorget and crown and the gray-green spotted undersides.

appear "dirty." The tail is rounded, with conspicuous white spots at the tips of the outer two pairs of feathers (r4 and r5). There is no rust in the tail.

JUVENILES: Resemble adult females. The grooves on the maxilla and the light tips to each crown and back feather will be lost a short time after hatching. The secondary wing feather tips are rounded.

JUVENILE MALE: There will usually be some large red spangles on the throat and often on the crown. As summer progresses, both the gorget and crown gradually fill with red spangles. Early in the year, they resemble adult females, but the outer tail feathers (r5) of juvenile males usually have black pigment, which fades into the white tip and/or a thin black line on the rachis in the white tip.

JUVENILE FEMALE: The chin, throat, and breast feathers have a brighter white background than those of the adult female. There are usually fewer (if any) red feathers on the throat than in adult

Anna's Hummingbird

Adult male, dorsal view.

Adult male, ventral view.

Adult female, dorsal view.

Adult female, ventral view.

Small Hummingbirds

Juvenile male, dorsal view.

Juvenile male, ventral view.

Juvenile female, dorsal view.

Juvenile female, ventral view.

Anna's Hummingbird

Adult male tail. While the central two pairs of feathers (r1 and r2) are green to match the back, the outer three pairs (r3–r5) are black, sometimes with a brownish tinge. The outer rectrix (r5) is narrow and sometimes curved. The tips are blunt on all feathers. Note that the central feathers are shorter than the outer feathers, giving the closed tail a notched appearance.

Adult female tail. As in most female hummingbirds, the outer pairs of feathers are dark with white tips. In Anna's, the feathers are much wider in relation to their length than those of other local species, except for Costa's and perhaps White-eared Hummingbirds. Note that no black central shaft is visible in the white tips of any of the outer feathers.

Small Hummingbirds

Juvenile male tail (right half). The general shape of the feathers is the same as that of either adult or juvenile females, but there is much less white at the tips of r4 and r5, and often there is no white at the tip of r3. Note the black feather shaft that extends into the white tip, not usually present in females.

Juvenile female tail. The shape of the feathers is the same as the shape of the juvenile male's tail, but there is more white at the tips of r4 and r5, and often there is some white at the tip of r3. Note that no black "bleeding" is present into the white tips.

Anna's Hummingbird

The adult female may have many red feathers in the center of the gorget, but they are smaller than those of the male.

Hovering adult male, showing the red gorget, spotted "dirty" flanks, and dark undertail coverts.

Small Hummingbirds

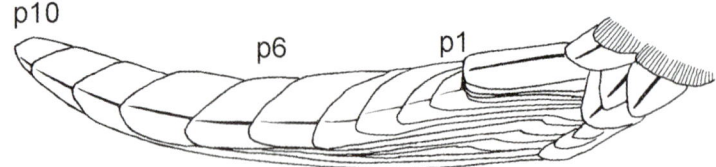

Wing of a female Anna's Hummingbird. Anna's and Costa's Hummingbirds are in the genus *Calypte*, and both have the same wing pattern. Compare with the *Archilochus* wing. Note that the widths of the anterior and posterior vanes of all of the primaries are about equal in *Calypte*, but not in *Archilochus*.

Female Anna's wing. Note the *Calypte* pattern of the wing, with the anterior and posterior vanes of each inner primary feather being about equal in width. Compare with the *Archilochus* wing.

Anna's Hummingbird

Juvenile male, showing large red spangles in the gorget and a few red feathers back of the eye.

Diagram of r5 of two juvenile male Anna's, with the black shaft in the white tip, or black bleeding into the white tip.

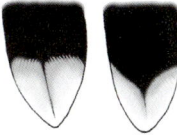

Diagram of the tips of two r3 feathers of juvenile Anna's. On the left is an r3 of a juvenile male, and on the right is an r3 of a juvenile female (with more white).

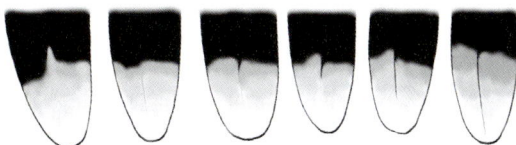

Diagram of the tips of r5 of female Anna's, either adult or juvenile. The black does not bleed into the white tip as much as in the juvenile male.

Small Hummingbirds

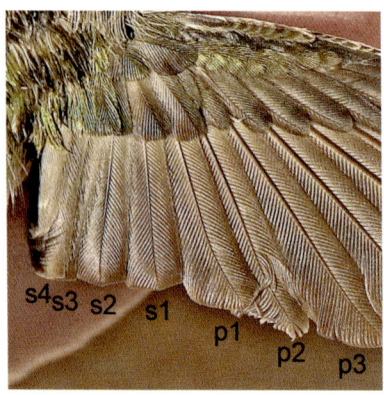

Secondary feathers (S) of an adult female Anna's. The vanes of the secondaries look as though they were snipped at the tip, making a broad V shape.

Secondary feathers of a juvenile female Anna's. The secondaries are rounded at the tip. This character can be used for most of the summer, but by fall, juveniles will have molted the secondaries to their first basic plumage, and these feathers will be snipped at the tip like those of the adult. At that time, the bird can no longer be determined to be a "hatching year" bird.

Anna's Hummingbird

Only a few older adult females have red spangles on the crown. They may also have more spangles than usual in the gorget. This bird is not a juvenile male—which might show a few spangles on the crown—as it has no pale fringes on other crown feathers and no grooves on the maxilla. The appearance of large spangles on the crown and in the gorget of some adult females may be related to hormonal changes.

This adult male appears to have a yellow forehead, but the yellow powder on the head and maxilla is in fact pollen. No North American hummingbird has a yellow crown or gorget.

Adult male in flight. The bright rosy red gorget and crown make identification easy. Note the shape of the tail feathers, parallel sided with a rounded tip—another good clue to this bird's identification.

Small Hummingbirds

females. R5 is wider than in the adult female. Black entering the white tip is lacking in both adult and juvenile females, but a thin black line may appear. See the variability in this trait in the figures on page 129.

SIMILAR SPECIES: Costa's males are smaller and have shorter bills and tails, and in good light they have a purple gorget and crown. Costa's females are also smaller, with shorter bills and tails and clearer, whiter undersides. Ruby-throated and Black-chinned females are thinner birds with longer bills. They also have a different primary feather shape, with a much narrower anterior vane than posterior vane of p1 through p6.

DISTRIBUTION: The Anna's Hummingbird is a bird of the Pacific coast, ranging from the Alaska panhandle south through coastal British Columbia and the Pacific coast of the United States, then spreading eastward into southern California, Arizona, southern New Mexico, and southern Texas, and south into northern Sonora and northern Baja California. Its breeding range is more limited, from extreme southern British Columbia to extreme northern Baja California and eastward into southwestern Texas. The breeding range is slowly extending northward, eastward, and into higher mountain elevations. Anna's Hummingbird is more tolerant of human presence, and it will nest in parks and gardens in urban settings. The availability of artificial nectar sources likely aids this range expansion.

MIGRATION: While some birds may be resident in southern areas, most migrate from the northern coastal regions to southern Arizona and northern Mexico to winter. In southeastern Arizona, there is a movement pattern in fall from west to east, with birds wintering from California to southern Texas, as far east as Houston.

COURTSHIP AND NESTING: Both sexes may defend feeding territories, which extend to one-quarter acre with a much larger buffer zone. The first males to arrive get the best territory: low flowering bushes surrounded by taller trees. Males defend their territory by flying, singing, and chasing off intruders. They prefer more open areas than do females. Because breeding occurs over a large latitudinal and altitudinal range, the breeding season is very protracted. It may start as early as December (following winter rains) in southern locations, and run through May at more northern ones. In courtship, the male makes long dives from as high as forty meters toward a female; hovers; then climbs in steps, showing his gorget and crown to the female at each step, until he reaches the top of the cycle. Then he dives again. At the end of the dive, the male makes a loud chirp with its tail feathers. Males also use a shuttle display. Females construct the nest, usually on a horizontal branch in oak (*Quercus* sp.), mesquite (*Prosopis* sp.), palo verde (*Cercidium* sp.), or numerous other trees and shrubs from one to nine meters above the ground. Nests are made of fine fibers such as cattail down, feathers, and flowers, stuck together with spiderweb. The outside is usually decorated with bits of lichen, moss, dead leaves, and other items that serve as camouflage. The two eggs are sometimes laid before the nest is complete. Incubation begins after the second egg is laid (forty-eight hours after the first egg) and hatching occurs in fourteen to nineteen days. Young birds are fed by the female for eighteen to twenty-six days in the nest and for up to two weeks postfledging. Young are fledged between February and May, depending on the latitude.

NUTRITION AND MOLT: Like other hummingbirds, Anna's depends on a good supply of nectar for its sucrose content. Plant species used for this include gooseberry (*Ribes* sp.), currant (*Ribes* sp.), manzanita (*Arctostaphylos* sp.), sage (*Salvia* sp.), penstemon (*Penstemon* sp.), fuchsia (*Fuchsia* sp.), columbine (*Aquilegia* sp.), and a wide variety of introduced plants in urban gardens. Anna's may also feed on sap and insects from natural seeps and woodpecker and sapsucker holes; large numbers of small flying insects; and insects gleaned from foliage. Adult molt starts in May and continues through November.

Small Hummingbirds

COSTA'S HUMMINGBIRD, COHU (*Calypte costae*) Colibrí de Costa (Sp.) *Calypte* = veiled, from the combined crown and gorget (Gould); *costae* = Louis Marie Pantaleon Costa (Bourcier & Mulsant)

IDENTIFICATION: Costa's Hummingbirds are smaller than their congener, Anna's Hummingbirds, and the males have bright purple gorgets and crowns instead of rosy pink. Costa's Hummingbirds can usually be identified by their posture. When perched, they often appear "neckless," with the head thrust forward. The bill is short and straight.

ADULT MALE: The crown and gorget are deep purple, but as with other species, the color is dependent on the angle of light. Sometimes it appears reddish violet. The gorget has "ears" of long feathers projecting along the sides of the neck. The iridescent color is only visible in good light; otherwise the crown and gorget appear black. There is a thin postocular white stripe. The bill is relatively short and straight. The breast is white. The flanks and

Adult male Costa's Hummingbird.

Costa's Hummingbird

Adult male, dorsal view.

Adult male, ventral view.

Adult female, dorsal view.

Adult female, ventral view.

Small Hummingbirds

Juvenile male, dorsal view.

Juvenile male, ventral view.

Juvenile female, dorsal view.

Juvenile female, ventral view.

Costa's Hummingbird

Adult male tail. Note the similarity to the adult male Anna's tail, with green central feathers and the rest dark, and with no white tips. Costa's tails are browner and the feathers are more curved—especially r5, which is also very thin (often thinner than in the tail shown here).

Adult female tail. Only the left half of the tail is shown. Note the large white tips on r3, r4, and r5. The feathers are relatively short and wide; they are similar to, but shorter than, those of Anna's females, showing the same general color pattern. Note that there is no black "bleeding" down into the white tip.

Small Hummingbirds

Juvenile male tail. Note that the feathers are short and relatively wide—similar to those of the adult female—but there is little to no white at the tip of r3. The black of the middle of the outer feathers bleeds into the white, and the black rachis extends into the white tip (see diagrams of Anna's tails).

Juvenile female tail. Only the left half is shown. Note the large white tips on r3, r4, and r5, almost identical to those of the adult female. The juvenile may have more white on r3, and the contrast between gray green, black, and white is stronger in the adult tail than in the juvenile tail. Again, no black bleeds into the white tips.

Costa's Hummingbird

Drawing of an adult male Costa's Hummingbird, emphasizing its plumage characters.

Adult male, showing the iridescent purple gorget and crown. Note the long "ears" that extend backward from the gorget, typical of Costa's.

Juvenile male, showing only a few purple spangles in the gorget on an otherwise brown face.

Small Hummingbirds

Juvenile male sleeping at night on a feeder. Note that the short tail does not extend beyond the tips of the folded wings. The bird looks chunky or stubby as a result.

Juvenile female, showing clear-white underparts, a short bill, and a tail not projecting beyond the primaries.

Adult female molting p9. P8 has grown in, and p10 is an old feather still in place. Note the unmarked underparts and short tail.

Costa's Hummingbird

Adult male in flight. The crown and gorget appear reddish purple in this photo, but normally they are a bluer purple.

sides of the belly are heavily mottled with gray and green, leaving a light stripe down the center. The tail is dark and short, with thin outer feathers (r5).

ADULT FEMALE: The back and crown are green. The undersides from chin to belly are very light and often white. The bill is short and straight. The short tail has white tips to the outer two or three pairs of feathers. Often there are a few small purple spangles on the throat. The black at the end of r5 does not enter the white tip, similar to the Anna's Hummingbird.

JUVENILES: Resemble adult females. The maxilla has fine corrugations, and each of the crown and back feathers has a buffy to whitish fringe at the tip for a short time after fledging.

JUVENILE MALE: There are usually some purple spangles on the throat, and rarely there are a few on the crown. The tail is short

Small Hummingbirds

and bronzy green, with black toward the ends of r3 through r5. There are white tips to r4 and r5 and sometimes r3. The white tip of r5 has black entering the white. Sometimes the black is simply a line in the center along the shaft, as in Anna's Hummingbird.

JUVENILE FEMALE: Very similar to the adult female. There are rarely any spots in the throat. The white tail spots will not have black entering the white.

SIMILAR SPECIES: The male Costa's is unlike any other hummingbird, but in poor light it might be confused with an adult male Anna's. Anna's is larger and heavier, with a longer tail and longer bill. A female Costa's could be confused with a female Anna's, Black-chinned, or Ruby-throated, and possibly a Calliope. Costa's is smaller than Anna's, with a shorter tail and bill. It is also not as dusky or "dirty" looking on the undersides and flanks. The female Calliope has an even shorter bill and tail, and the flanks are tinged with cinnamon. Female Anna's typically have a center gorget spot; female Costa's typically do not. Black-chinned and Ruby-throated females are thinner and have longer bills and tails. They also tend to have some faint buffy color on the flanks.

DISTRIBUTION: The Costa's Hummingbird is a bird of the desert. It breeds as far north as southern Nevada and Utah, and south through western Arizona into Sonora, Mexico. It is a permanent resident of the deserts of southwestern Arizona, western Sonora, and southern and Baja California, down the Sierra Madre Occidental to Sinaloa and, in winter, south along the coast to Jalisco. In May and June, the Costa's moves to higher elevations. It appears to leave the desert during the hottest and driest time of the year. Birds may wander east through Texas and north to southern Alaska. Habitat for the Costa's is decreasing rapidly along the California coast and throughout Mexico, where desert scrub is being cleared and replaced with grass for cattle grazing.

MIGRATION: For many birds, there is no long-distance migration, as they are already on location and ready to breed at the earliest opportunity. In California, birds move north to the breeding grounds as early as January, and they depart by late April or early May. In western Arizona, adults move west in March and return with young birds in late June. Many overwinter, or they may be permanent residents in southeastern Arizona at low elevations.

COURTSHIP AND NESTING: Nesting begins for most individuals by mid-February and may run through mid-April, with young fledged from March to mid-May. Their nesting cycle is dependent on nectar availability, which in turn is dependent on the flowering of desert shrubs. That, in turn, depends on the abundance of winter rainfall. Second broods are not common in the Arizona population, which breeds later than those in California. Some believe that the birds depart from their desert habitat after raising one brood, and that they start a second nesting attempt in the chaparral at higher elevations later in the spring. The nest is a 1½-inch diameter cup made of down, leaves, feathers, and flowers, and it is bound with fine fibers and spiderwebs. It is built on the small branches and leaves of desert plants, including ocotillo (*Fouquieria splendens*), century plant (*Agave* sp.), cholla (*Opuntia* sp.), jojoba (*Simmondsia chinensis*), mesquite (*Prosopis* sp.), acacia (*Acacia* sp.), deer brush (*Ceanothus* sp.), and buckthorn (*Rhamnus* sp.), usually only one meter above the ground. In some areas, they show a strong preference for palo verde (*Cercidium* sp.) trees in washes and side hills, where nests may be two meters above ground level. In other places, they may build on ragweed plants (*Ambrosia* sp.), which are only a half-meter over the ground. Nests are smaller and more loosely constructed than those of other similarly sized species, perhaps due to the urge to complete nesting before resources dry out and nectar-producing flowers are gone. Females incubate two eggs for fifteen to eighteen days and brood an additional twelve days after hatching. The young fledge from twenty to twenty-three days after hatching.

NUTRITION AND MOLT: Like other small hummingbirds, Costa's rely primarily on nectar for their energy source, which may be hard to come by at times in the desert. Like other hummingbirds, the Costa's protein and fat are derived from insects and arachnids, caught by "fly-catching" or plucked from leaves and branches. Adults have already started body and primary molt by the time we see them in June and July, often at higher elevations. Molt may start in late June and continue through the winter; all body and flight feathers are replaced.

CALLIOPE HUMMINGBIRD, CAHU (*Selasphorus calliope*) Colibrí de Caliope (Sp.) *Selasphorus* = moving or carrying light (Swainson); *calliope* = the muse of epic poetry, also beautiful voice (Gould)

IDENTIFICATION: The Calliope Hummingbird is the smallest breeding bird in North America, weighing only about 2.5 g, the weight of a modern US penny. When perched, its wings are longer than the tail. The central pair of rectrices (r1) is spatulate in all ages and both sexes, but this is most obvious in the adult male.

ADULT MALE: Males are easily recognized by their dark rose-red gorget feathers, which are arranged in long lines that extend down the throat and farther down the sides of the neck. The crown and back are olive to golden green. The breast, belly, vent band, and undertail coverts are white. The tail is pale gray, with some rufous along the feather edges toward the base. The bill is thin, short, and straight.

Adult male.

Calliope Hummingbird

Adult male Calliope Hummingbird, dorsal view.

Adult male, ventral view.

Adult female, dorsal view.

Adult female, ventral view.

Small Hummingbirds

Juvenile male, dorsal view.

Juvenile male, ventral view.

Juvenile female, dorsal view.

Juvenile female, ventral view.

Calliope Hummingbird

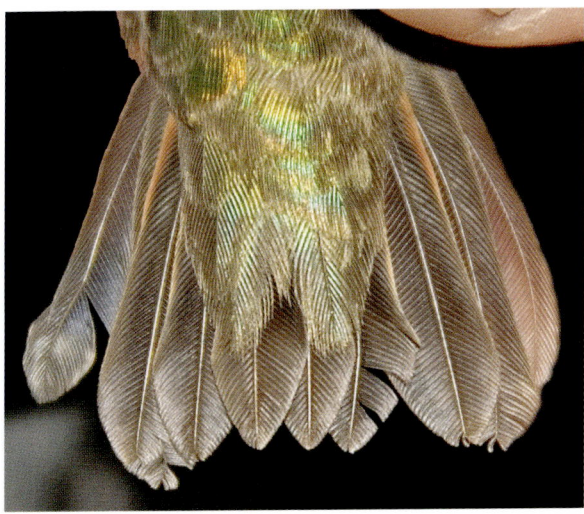

Adult male tail. The very short tail has all-brown feathers, with some rusty color on the margin of r3. Both r1s show their spatulate shape, characteristic of Calliope Hummingbirds.

Adult female tail. The short tail has some rusty color at the base of most of the outer pairs of feathers, and the rufous color may extend down the feather margins or along the feather shafts. Rectrices 3, 4, and 5 have white tips. Often the white on r3 is worn away.

Small Hummingbirds

Juvenile male tail. The right side of the tail is fanned open to see the rusty margins of r1 through r4. Again, note the spatulate shape of both r1s. The white tail tips are similar to those of the adult female. Some juvenile males have a small white tip on r2, similar to the juvenile female.

Juvenile female tail. It is similar to the juvenile male tail, but it lacks the rufous margins to r3 and r4. There may be some slight rusty coloring to the base of those feathers. The amount of white at the feather tips is about the same as in the juvenile male, except that the juvenile female may also have white on r2. The central pair of tail feathers (r1) does not have a well-defined spatulate shape in either the adult or juvenile female, although sometimes there is a hint of such. In all ages and both sexes, the tail is very short.

Calliope Hummingbird

Adult male Calliope Hummingbird, as seen in the field. The male is easy to identify by its rows of long, thin spangles that project away from the body.

Adult female Calliope Hummingbird, as seen in the field. The female is more difficult to identify, but its very short bill, dumpy or humpbacked posture, short tail, clear breast and belly, and rusty flanks should help. You can see the rusty margin to r5 as well.

Wing of a Calliope Hummingbird. The feather pattern is more like that of *Calypte*, with almost equal-width anterior and posterior vanes on the inner primaries. We know of no wing feather patterns that can help with age or sex determination.

Small Hummingbirds

These adult males show how the spangle feathers are unlike those of other North American hummingbirds, as they are long, thin, and arranged in rows. When displaying, the bird can elevate and spread the feathers, as seen in the right photo.

The adult female's throat has rows of dark spots, with occasional small red spangles.

The female head resembles that of Rufous, Allen's, and Broad-tailed Hummingbird females. The bill is much shorter, however, and rows of dark spots are evident. The face is more gray green than any of the other *Selasphorus* hummingbirds.

Calliope Hummingbird

The juvenile male is a browner version of the adult female, with more rusty color in the plumage. Lines of dark feathers are evident on the gorget. The corrugations on the bill and the tail pattern are better clues to the age and sex of juveniles.

Adult male in flight. With its series of long lines of spangles, the gorget is unique. The tail is very short.

ADULT FEMALE: The crown and back are olive to golden green. There may be a few red feathers on the throat and many dark spots. The flanks are a pale rust. There are white tips on the outer two, and sometimes three, tail feathers. The shape of r1 is less spatulate than in the adult male. There is some rufous on the margins to the base of some tail feathers.

JUVENILES: Resemble adult females. For a short time after fledging, there are corrugations along the maxilla and light fringes on the crown and back feathers.

JUVENILE MALE: Young males usually have a few to many red spangles in the gorget. The base of some of the tail feathers has rufous along the feather edges.

JUVENILE FEMALE: There are white tips on the three outer tail feathers (r3–r5) and sometimes on the second rectrix (r2). The base of the tail feathers may have a pale rufous cast, but there is no dark rufous along the feather margins as in juvenile males and adult females. There are no red feathers on the throat, only light spotting.

SIMILAR SPECIES: All other hummingbirds are larger and have longer bills than the Calliope, except for the accidental Bumblebee Hummingbird, which has a shorter bill. The Bumblebee's gorget is full feathered and more wine colored. The Bumblebee also has much more rufous in the tail. The female Costa's has a grayish-green wash on the flanks, while the Calliope's flanks are rusty.

DISTRIBUTION: Calliope Hummingbirds are abundant breeders in the northern Rocky and Cascade Mountains, from northeastern California, western and central Nevada, central Utah, northwestern Wyoming, the Cascades of Oregon and Washington, north and central Idaho, the Rockies of western Montana, Alberta, and southern to central British Columbia. They winter in the Sierra Madre Occidental from Durango, Nayarit, and Jalisco south to Oaxaca, and they rarely winter in southern Texas. Calliopes breed most abundantly at high elevations, nesting in shrubs and saplings that have sprouted following logging or fire.

MIGRATION: In between their breeding and wintering grounds, there are several thousand miles for this tiny bird to negotiate back and forth each year. Little is known about the migration routes of the species. Migration begins in March and continues through May. Males precede females, often arriving before nectar plants have flowered. In migration they are seen at lower elevations, and they seem to appear wherever there are good stands of nectar-producing flowers. It is thought that they move north in April and May along lower-elevation areas of the Pacific coast, and that they return south more to the east, in the higher elevations of the Rocky Mountains. They are rarely seen in Arizona in spring, but are more commonly encountered in the fall.

COURTSHIP AND NESTING: The nest is built on a single branch, usually under another branch or leaf that serves to protect it from weather and decrease the radiation of its body heat to the cold night sky. Nests are from two to twelve meters above ground, and they can be in any young conifer or other tree, e.g., pine (*Pinus* sp.), fir (*Abies* sp.), Douglas fir (*Pseudotsuga menziesii*), apple (*Malus domestica*), or alder (*Alnus* sp.). Nests may be reused the following year, or at least some of the material is recycled. The two eggs are incubated by the female for fifteen to sixteen days, and the young fledge after another eighteen to twenty-one days.

NUTRITION AND MOLT: Foods for adults and young are nectar from flowers and occasionally sapsucker wells, and small insects, which are caught on the wing when the bird darts out from an exposed perch. There is no data to detail the sequence and extent of molts for this species.

BROAD-TAILED HUMMINGBIRD, BTLH (*Selasphorus platycercus*) Zumbador Coliancho (Sp.) *Selasphorus* = moving or carrying light (Swainson); *platycercus* = flat or broad tail (Swainson)

Small Hummingbirds

A Colorado Broad-tailed Hummingbird had the longest documented life span of any hummingbird of any species—over twelve years.

IDENTIFICATION: Broad-tailed Hummingbirds are larger or longer and have much larger tails than most of the other small hummingbirds.

ADULT MALE: The crown and back are brilliant emerald to golden green. The upper tail coverts are golden changing to more olive green. The gorget is bright rose red. The upper breast, center line of belly, and vent band are white. The flanks are green mottled with darker green, gray, and some rust. The undertail coverts are grayish. The tail is proportionately long and wide, with green central feathers (r1). The other tail feathers are black with rusty margins. The central pair (r1), r2, and r3 taper to a point. Primaries 9 and 10 are modified to create a trilling sound (see page 160).

Adult male Broad-tailed Hummingbird.

Broad-tailed Hummingbird

Adult male, dorsal view.

Adult male, ventral view.

Adult female, dorsal view.

Adult female, ventral view.

Small Hummingbirds

Juvenile male, dorsal view.

Juvenile male, ventral view.

Juvenile female, dorsal view.

Juvenile female, ventral view.

Broad-tailed Hummingbird

Adult male tail. R2 through r5 are all dark, with no white tips. Sometimes there is white at the tip of r5. Both r2 and r3 are dull pointed, and r2 and usually r3 have rusty margins at the base.

Adult female tail. R1 is bronze green, and r2 is green with a dull-pointed black tip. Note the rusty base to r3 through r5, with no black in the rusty area, followed down the feather by some green and black and a white tip.

Small Hummingbirds

Adult female tail (right half). Here the black tip and green base of r2 are clearly visible. Sometimes there is a small white tip to r2. Note that there is some rufous along the basal margin of r2. The rufous, green, black, and white bands to r3 and r4 are clearly shown.

The juvenile male tail color pattern is very similar to that of the adult female, but the feather tips are more pointed, and the black "bleeds" up into the basal rufous color along the rachis.

Broad-tailed Hummingbird

Juvenile male tail (left half). Note the more pointed r2, which is mostly green but with a black tip and rufous margin toward the base of the feather. In this bird, the green—rather than the black—"bleeds" up along the feather shaft into the rufous band of color.

Juvenile female tail (left half). The color pattern is almost identical to that of the adult female. The tips of all feathers are more rounded than in a juvenile male. Without grooves in the bill to indicate that this is a juvenile bird, it would be impossible to determine the age only from the tail pattern or shape.

Small Hummingbirds

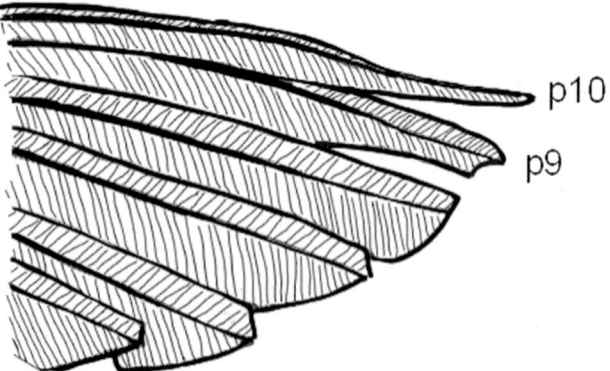

Photograph and diagram of the wing tip of an adult male Broad-tailed Hummingbird. The pointed shape of p10 and the cut-off shape of the end of p9 may account for the trilling sound that adult males make when they fly. However, they can fly without making the trill.

Broad-tailed Hummingbird

Gorget of an adult male. The rose-red gorget and green cap are distinctive. Note that there is no black on the chin or around the eyes, as there would be on an adult male Ruby-throated Hummingbird.

Gorget of an adult female. Females sometimes have a few red spangles on the throat, and they always have a few rows of dark (often green) spots. Juveniles also have spots.

Wing of a second-year male in May, showing that p10 is almost grown in and p9 is only halfway in.

Small Hummingbirds

Head of a juvenile male. Note that the male, with its green gorget spots, has more green than the female in the next images.

Heads of two female Broad-taileds. An adult is on the top and a juvenile is on the bottom. Note that both females are browner on the head than the male, and they have brown instead of green spots on the throat. The female face pattern lacks much rusty color. Compare with Rufous Hummingbird females.

Broad-tailed Hummingbird

Adult female in the field. Compared to other *Selasphorus* hummingbirds, the Broad-tailed is very green on the back and usually on the head. Note the rusty flanks, the long tail with white tips, and the rusty margin to r5.

This adult female has a bright green back and spotting on the throat. The angle of the bird in relation to the camera makes the bird and its tail appear shorter than they really are. The bill is too long for either a Calliope or Costa's Hummingbird.

Small Hummingbirds

Adult male in flight. The bright red gorget reminds you of a Ruby-throated Hummingbird, but this bird lacks the black chin and face, and it has rusty color along the margin of the outer tail feathers.

ADULT FEMALE: The crown, back, and upper tail coverts are golden to olive green. The throat is pale with scattered gray or bronze spots. The breast, belly, and vent band are white. The undertail coverts are rusty. The tail is large, with each feather blunt at the tip and rusty at the base. There is black before the white tips of r3 through r5. The central pair (r1) is bronzy green without white tips, and it may have rufous along the margins. R2 is bronzy green, with a dark band toward the tip that is usually without any white. Often there is green and black smudging into the rufous of r3, r4, and/or r5.

JUVENILES: Resemble adult females. The corrugations along the maxilla and the beige tips to the crown and back feathers persist for a short time after fledging.

JUVENILE MALE: Males often show some red spangles on the throat. Many males do not develop the modified p9 and p10 until spring of the year after hatching. R1 is all green, sometimes with rufous margins. There is no white at the tip of r2. Juvenile males show the black and/or green of r3, r4 and r5, which "bleeds" up or smudges into the rufous base along the feathers' rachis. Male tail feathers are more sharply pointed than female rectrices, which are rounded.

JUVENILE FEMALE: Juvenile females are very similar to adult females. R1 is green with a black tip, and it lacks rufous on the margins. The tip of r2 is rounded and may have some white. R3, r4, and r5 may have some green smudging into the rufous at their base.

SOUNDS: Adult males have highly modified ninth and tenth primary wing feather tips. This creates a high-pitched trill when the wings beat rapidly, which is unique to this species in North America.

SIMILAR SPECIES: A person coming from the eastern United States might immediately think that Broad-tailed males are their familiar Ruby-throated Hummingbirds, as both are "green" with a red gorget. But male Broad-taileds have a rosier red gorget without the black chin, as well as a more brilliant green back. The tail is also very different than the forked tail of the Ruby-throated. Females resemble the smaller female Rufous and Allen's Hummingbirds, which have shorter bills, wings, and tails and more rufous on the undersides. The wing chord of the Broad-tailed is 46.5 mm or longer. Rufous and Allen's wing chords are shorter. The width of r5 (over 4.3 mm in females and over 3.1 mm in young males) is greater than that of either Rufous or Allen's Hummingbirds.

DISTRIBUTION: Broad-tailed Hummingbirds breed in the higher-elevation pine (*Pinus* sp.), fir (*Abies* sp.), and Douglas fir (*Pseudotsuga menziesii*) forests of the United States and Mexico. Therefore, their range is scattered and disjunct, with populations in far eastern California, southern Idaho, Colorado, Utah, Wyoming, Nevada, Arizona, New Mexico, and Texas, as well as along the crests of the Sierra Madres and the highlands of central and southern Mexico. They winter in the highlands of northern Mexico south to western Veracruz and Oaxaca, and south to Guatemala. Like other western hummingbirds, in winter Broad-taileds may be found across the southern states, from Texas east.

MIGRATION: Little is known of the migratory pathways of Broad-taileds, which come north out of Mexico in spring and return in fall. Males arrive on territory before females, from early March (southeastern Arizona) to late May (Idaho); they arrive in Colorado in late April and early May. For birds departing the Colorado area of the Rocky Mountains, fall migration begins the first of August and continues through the end of September. Birds leave Arizona in late August through early October.

COURTSHIP AND NESTING: Broad-taileds nest in the coniferous forest ecosystem, in trees that include pine, juniper (*Juniperus* sp.), oak (*Quercus* sp.), fir, Douglas fir, spruce (*Picea* sp.), and aspen (*Populus* sp.). They also nest along riparian areas of willow (*Salix* sp.) and mountain mahogany (*Cercocarpus ledifolius*). Nests are built within one meter of the ground in a sheltered area with protective vegetation overhead, which helps reduce heat loss during cold nights. Egg laying begins in late May, and the last eggs are laid by the beginning of July. Young appear in late June and are seen through mid-August. Incubation by females only lasts sixteen to nineteen days; it begins with the laying of the first egg. The first egg usually hatches one day before the second. Birds are fed in the nest for about twenty days, and for several days more after fledging. The length of the incubation period can be increased by several days if rain or extreme cold prevents the female from feeding. If these weather conditions occur, the female then enters torpor and the egg temperature drops, reducing the developmental rate of the embryo.

NUTRITION AND MOLT: Food consists of nectar from a large number of alpine flowers, e.g., columbine (*Aquilegia* sp.), scarlet gilia (*Ipomopsis aggregata*), penstemon (*Penstemon* sp.), and paintbrush (*Castilleja* sp.), as well as small insects gleaned from foliage and caught while hovering and flying. Small flying insects are the principal diet item for young birds, mixed with nectar that the female pumps into the young's stomach. Birds molt in winter in Mexico, but often many first-year males return with the juvenile p9 and p10 not yet replaced.

RUFOUS HUMMINGBIRD, RUHU (*Selasphorus rufus*) Zumbador Rufo (Sp.) *Selasphorus* = moving or carrying light (Swainson); *rufus* = reddish (Gmelin)

IDENTIFICATION: Rufous Hummingbirds are the most aggressive hummingbirds in North America. Rufous and Allen's Hummingbirds are very difficult to separate in the field where and when ranges overlap (see Similar Species on page 177).

ADULT MALE: The crown is dark green, the back is reddish brown, and the upper tail coverts are also rusty. Some males are solid rufous on the back, while others have varying numbers of green feathers, usually in the center of the back. The gorget is brilliant orange red. The breast, middle belly, and vent band are white. The flanks and undertail coverts are rusty. The tail is dark rust, with a blackish pointed tip to each feather. Almost all adult males have a distinct notch on the inner (medial) web at the tip of r2.

ADULT FEMALE: The crown and back are green. There are rusty tips to the upper tail coverts, and the flanks and undertail coverts are rusty. Bronzy-green spots appear on the throat, often with a few to many small red-orange spangles in the center. The breast, belly, and vent band are white. The central tail feathers (r1) are green, with rust at the base and black at the tip. The basal half of the other tail feathers is rusty; the distal half is black. There is usually a dent in the inner and/or outer web of r2. The rounded tail tips of r3, r4, and r5 are white. The width of r5 where the black meets the white is greater than 2.7 mm.

Small Hummingbirds

Adult male Rufous Hummingbird.

JUVENILES: Resemble adult females. The thin, pale cinnamon fringes on the crown and back feathers are present for only a short time after fledging, and most wear away before the birds arrive in Arizona. The maxilla has fine corrugations, which disappear as the bill hardens.

JUVENILE MALE: Usually there are one or more large orange-red spangles scattered on the throat. The rufous base of r1 occupies more than half the length of the feather. R2 almost always has a dent in the lateral margin at the tip. The width of r5 is greater than 2.6 mm.

JUVENILE FEMALE: Resembles the adult female. There are few or no spots on the throat and usually no orange spangles. The rufous at the base of r1 is usually less than half the length of that feather. There is usually no dent in the tip of r2. The width of r5 is greater than 3.3 mm.

Rufous Hummingbird

Adult male, dorsal view.

Adult male, ventral view.

Adult female, dorsal view.

Adult female, ventral view.

Small Hummingbirds

Juvenile male, dorsal view.

Juvenile male, ventral view.

Juvenile female, dorsal view.

Juvenile female, ventral view.

Rufous Hummingbird

Adult male Rufous Hummingbird tail (right half). The tail is all rufous and black. Note the shape of the tip of r2, with a notch in its inner or medial margin and a dent in the outer or lateral margin. This pattern, shown here in an extreme form, is present in almost all adult male Rufous Hummingbirds. Note that r5 is very narrow, but compare it with the r5 of an Allen's Hummingbird, which is almost like a needle.

Adult female tail (left half). Rectrices 2 through 5 have green below the rufous, then black, and finally, on r3 through r5 and usually on the tip of r2, white. Note the very slight "dent" to the medial side of the tip of r2, reminiscent of the notch in adult male r2 feathers.

Small Hummingbirds

Tail of a juvenile male. All of the feathers are pointed at the tip. The rufous at the base covers more than half of each feather. There is a dent in the lateral tip of r2, and r5 is relatively narrow.

Tail of a juvenile female. The feather tips are more rounded than in a male. The white tips are larger. There is little rusty color in r1—only at the base—and the black sections are longer than in the juvenile male's tail.

Rufous Hummingbird

Gorget of an adult male, showing the typical orange-red spangles. Note how, at a different sun angle, they appear yellow on the margin.

Gorget of an adult female, showing the mostly unmarked throat, except for a few scattered spangles.

Gorget of a juvenile male, which has many lines of dark brown spots and a few spangles.

Small Hummingbirds

Head of an adult male. The gorget appears almost black out of the light, but the rufous color of the head indicates that it is either a Rufous or Allen's.

Adult male Rufous. Just because there are green feathers on the back does NOT mean this bird is an Allen's. You have to look at the tail!

Rufous Hummingbird

Head of an adult female. Note the light spotting and thin cinnamon eyebrow.

Head of a juvenile male. Note the heavy spotting and cinnamon over the eye.

Gorget of a juvenile female. It has no or very few spots and a few developing spangles.

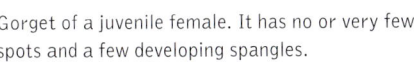

Small Hummingbirds

Head of a juvenile female. Note the cinnamon eyebrow and dark spots.

On this adult male, look for the notch in the medial side and a dent on the lateral side of the tip of r2.

Adult female Rufous Hummingbird. The smaller spangles are usually only in the center of the gorget

Juvenile male Rufous Hummingbird. Over half of each tail feather is rufous.

SIMILAR SPECIES: Rufous and Allen's adult males are very similar. Both are smaller than Broad-tailed Hummingbirds. The wing chords are less than 46.0 mm.

Although most Rufous males have almost all-rusty backs, if the back is mostly green, the bird could be either a Rufous or an Allen's. Look for a notch on the medial vane of r2; if present, the bird is probably a Rufous. If there is a slight emargination, it could be either a Rufous or an Allen's. The Rufous's tail is not as pointed as Allen's, and the outer tail feather (r5) is broader (1.8–2.6 mm) than Allen's (1.1–1.9 mm). The difference is only tenths of a millimeter, but the very thin outer tail feather of most Allen's Hummingbirds is like a needle, and it is often easy to see in the field.

Adult female Rufous and Allen's Hummingbirds look the same and are practically impossible to differentiate in the field. The tail feather tips of both are rounded. There is a slight emargination on the inner web of r2 in the female Rufous, which is very slight or absent in the female Allen's. Female Allen's tail feathers appear more pointed than those of the Rufous. The width of r5 is greater than 2.7 mm in the female Rufous and less than 2.7 mm in the female Allen's. Both characters are very hard to see unless the bird is in hand.

The tail feathers of both juvenile male Rufous and Allen's have rufous at the base and end in a black band, with a white tip on r3 through r5. There may be a slight indentation or a notch in r2 in the juvenile male Rufous, but not in Allen's. The width of r5 is greater in the Rufous (2.7–3.7 mm) than in the Allen's (less than 2.7 mm), but it takes a caliper to make the measurement. Juvenile female Rufous and Allen's appear almost the same, except that there may be a faint indentation on r2 in the Rufous that is missing in Allen's, and r5 is wider in Rufous (over 3.3 mm) than Allen's (less than 3.3 mm). Adult females and juveniles usually have a cinnamon eyebrow, which is lacking in the larger Broad-tailed Hummingbirds. The wing chord of Rufous and Allen's is almost always less than 46.0 mm.

DISTRIBUTION: The Rufous Hummingbird breeds farther north than any other hummingbird species. It prefers the inland mountain fir (*Abies* sp.), Douglas fir (*Pseudotsuga menziesii*), and spruce (*Picea* sp.) forests, from extreme northern California and western Oregon north through Washington, western Montana, the Rocky Mountains in Alberta and British Columbia, and southern Yukon Territory. It also prefers the Pacific coastal rain forests of Douglas fir and Sitka spruce (*Picea sitchensis*) of northern British Columbia and southeastern and south-central Alaska. Its wintering grounds—in west coastal and south-central Mexico, and along the Gulf Coast from Texas to Florida—are some 1,500 km from their southernmost breeding grounds. In between these two areas, the birds only appear in spring and fall as migrants. In Mexico, wintering birds prefer habitats with oak mixed with conifers.

MIGRATION: The widely accepted route of migration for Rufous Hummingbirds is north in the spring along the Pacific coast of Mexico, the United States, and Canada, and south in the fall through the Pacific Coast Ranges and Sierra Nevada and Rocky Mountain states, back to Mexico. Most birds may make this loop, but not all do, as we find many adults moving north in spring through southeastern Arizona, east of the primary route. The spring migration is protracted, with some individuals moving north in February and others not moving north until April. Males precede females to the breeding grounds. Most arrive at least by late April and early May, and they immediately nest. Fall migration begins in late June for adult males, and it extends into August for females and juveniles, with juveniles reaching southeastern Arizona in late September and early October in most years. This species remains aggressive throughout migration, as well as on the wintering ground where it is locally resident for up to one month. Each year, some individuals move east and are sighted in many of the southeastern and Atlantic coast states in winter.

COURTSHIP AND NESTING: Nests built early in spring are from one to three meters above ground, often in spruce, fir, or Alaska cedar (*Callitropsis nootkatensis*) trees. Those built later in the summer are often in birch (*Betula* sp.), hemlock (*Tsuga* sp.), or maple (*Acer* sp.), and they are usually higher, up to nine meters above the ground. Nests are constructed of downy plant parts, woven and stuck together with spider webbing. The outside surface is covered with lichens, bark, leaves, and bits of moss stuck on with spider silk. Nests may be renovated and used in subsequent years. Two eggs are laid, and young appear from the first of May through the first of July. Incubation lasts fifteen to seventeen days, and fledging does not occur until at least twenty days after hatching. (More information is needed.)

NUTRITION AND MOLT: The Rufous Hummingbird is known as the most aggressive small hummingbird in North America. Birds jealously guard feeders and drive all other birds away. Food primarily consists of nectar from tubular flowered plants, e.g., columbine (*Aquilegia* sp.), penstemon (*Penstemon* sp.), hummingbird trumpet (*Ipomopsis* sp.), and Indian paintbrush (*Castilleja* sp.), as well as small insects caught in flight or gleaned from foliage. They are important pollinators of plants. Because of the small size and long migratory flight of the Rufous, its metabolism and physiology have been studied more than those of any other hummingbird species. Their metabolism increases two and a half times, with an ambient temperature drop from 39°C to 20°C. They are also capable of torpor, allowing their body temperature to drop from a normal 39°C to as low as 12°C to conserve fuel stores. They can accumulate fat for migration, raising their body weight from around 3.5 g to as much as 5.7 g. Molt occurs in winter.

ALLEN'S HUMMINGBIRD, ALHU (*Selasphorus sasin*) Zumbador de Allen (Sp.) *Selasphorus* = moving or carrying light (Swainson); *sasin* = from a note by Captain James Cook in a French edition of the account of his third voyage (Lesson)

The Allen's Hummingbird was named for Charles A. Allen, who first noted the differences between Rufous Hummingbirds and this species (see Similar Species on page 186).

Small Hummingbirds

IDENTIFICATION: A small, aggressive hummingbird, Allen's are often indistinguishable from Rufous Hummingbirds in the field.

ADULT MALE: The crown and midback are emerald to olive to bright green. The upper tail coverts, undertail coverts, and flanks are rusty. The gorget is bright orange red. The upper breast and belly, including the vent band, are white. The tail feathers are very narrow and sharp pointed, with rust at the base and black at the tip. There is no notch in r2, but there may be a hint of emargination. The outer rectrix (r5) is needlelike and less than 1.9 mm wide where the black and rufous meet.

ADULT FEMALE: The crown and back are olive to golden green. The upper tail coverts are mostly green, with rusty tips. The throat may have a few to many small orange-red spangles. The breast, belly, and vent band are white. The flanks and undertail coverts are rusty. The central tail feathers (r1) are green, with rust at the base and a black tip. The other feathers are rusty at the base and black toward the tip. The rusty color occupies half or less of the total feathers' length. There are white tips on r3, r4, and r5. Rarely is there any emargination on the medial side of r2, and there is rarely a slight dent in the outer web. The width of r5 is less than 2.8 mm.

JUVENILES: Resemble adult females. There are corrugations on the maxilla for a short time postfledging. The back and crown feathers have buffy fringes, which quickly wear off.

JUVENILE MALE: There are often some large orange-red spangles scattered on the throat. The outer rectrix (r5) is narrower (less than 2.7 mm) than in juvenile females (less than 3.3 mm).

JUVENILE FEMALE: Usually there are no orange-red spangles on the throat, and sometimes there are no bronze spots either. The outer tail feathers are wider than those of the adult female but narrower than those of Rufous juvenile females.

Allen's Hummingbird

Adult male Allen's Hummingbird, dorsal view.

Adult male, ventral view.

Adult female, dorsal view.

Adult female, ventral view.

Small Hummingbirds

Juvenile male, dorsal view.

Juvenile male, ventral view.

Juvenile female, dorsal view.

Juvenile female, ventral view.

Allen's Hummingbird

Adult male Allen's Hummingbird tail (right half). Compare to the Rufous Hummingbird, and note the similarity in color pattern and general shape. Allen's tail feathers are more pointed and narrower than those of the Rufous. There is no notch and only a slight "dent" at the tip of r2, and r5 is narrower than any Rufous r5.

Adult female tail (right half). Compare to the female Rufous Hummingbird; note the similarity in color pattern and shape. There is a slight hint of a "dent" in r2, which is only present in a low percentage of adult female Allen's. The width of r5 where the black meets the white is 2.8 mm or less. Because this measurement overlaps with the narrowest r5 measurements in Rufous adult females (2.7 mm), sometimes these two cannot be accurately separated.

Small Hummingbirds

Juvenile male tail. Compare with the Rufous Hummingbird. The tail feathers of the Allen's are much narrower than those of the Rufous, and there is no hint of a notch or "dent" at the tip of r2. The width of r5 where the black meets the white is less than 2.7 mm. This just overlaps with the narrowest measurement of r5 in Rufous juvenile males; sometimes the two birds cannot be distinguished on that character alone. It is practically impossible to determine species in the field with certainty. You must be able to get a good look at the tail and, usually, make a measurement of r5.

Juvenile female tail. In almost all characters, this tail cannot be distinguished from a juvenile female Rufous Hummingbird's in the field. Note that r2 through r5 are more pointed than in the Rufous, and there is no hint of a notch or "dent" at the tip of r2. The width of r5 where the black meets the white is always less than 3.3 mm. In Rufous juvenile females, that width is always greater than 3.3 mm.

Allen's Hummingbird

Gorget and head of an adult male Allen's, which are practically indistinguishable from those of a Rufous. The color of the gorget seems pinker in this photo, but slight changes in light angle and intensity can vary the color.

Head of an adult male Allen's, practically indistinguishable from that of a Rufous.

Adult male in flight. This might be either a Rufous or an Allen's Hummingbird, because you cannot see the green back and the very narrow outer tail feather (r5). You have to check the tail of any Rufous/Allen's with green on its back to see if there is a notch and dent in r2; if there is not, check to see if r5 is very narrow.

SIMILAR SPECIES: All ages and both sexes of Allen's Hummingbirds are easily confused with Rufous Hummingbirds. Both have a compact shape, buffy-orange flanks, and varying amounts of rusty red brown in the tail. Allen's are, on average, slightly smaller than Rufous. For example, the wing chord of an adult male Allen's is 36.0–40.0 mm, and that of an adult male Rufous is 38.0–42.0 mm. Adult males have green backs, while most Rufous have rusty backs (although some will have a few to many green feathers). Rufous males have a distinct notch in the medial side of r2, missing in Allen's. The Rufous outer rectrix (r5) is wider (1.8–2.6 mm) than Allen's (1.1–1.9 mm).

Adult females have the same plumage as adult female Rufous, and they cannot safely be separated from Rufous in the field. Usually there is no emargination of the inner web of r2, and r5 is narrower (1.8–2.8 mm) than in Rufous (2.7–4.0 mm; measured where the white tip meets the black above it). The female may be confused with the female Bumblebee Hummingbird, which is smaller, has a shorter bill, and has much less rust in the tail. The Bumblebee has a buffy terminal spot on r2, while the Allen's r2 is all black. The Bumblebee is quiet and not as aggressive as the Allen's, and it often forages near the ground.

DISTRIBUTION: The Allen's Hummingbird has the smallest breeding range of any North American hummingbird. It is common along the coast of California, from about San Diego north to southern coastal Oregon. The breeding ranges of Allen's and Rufous Hummingbirds do not overlap, making it easier to separate these two species in summer. The Rufous breeds in the mountains and coasts farther north. There are two distinct subspecies populations of Allen's Hummingbirds: one (*S. s. sasin*) is a migrant to south-central Mexico, and the other (*S. s. sedentarius*) is a resident on six of California's offshore Channel Islands, and from coastal Los Angeles to extreme northern San Diego.

The wintering range of migrants is a narrow area in the mountains of central Mexico; little is known about the extent of this range. In fall migration, some birds come east into southeastern

Allen's Hummingbird

Arizona, where banding results show that less than 10 percent of juvenile birds known to be either Rufous or Allen's are in fact Allen's. The Allen's has also been found in southern New Mexico, across Texas, along the Gulf of Mexico, and throughout some of the southeastern states in winter. It shows up rarely along the Atlantic coast in winter.

MIGRATION: Migration north is along the Pacific Slope out of Mexico. Males arrive on the breeding grounds sometime in January and February, ahead of females. Adults migrate south first, followed by juveniles, on an inland route at higher elevations in the Sierra Nevada and coastal mountain ranges of California into Mexico.

COURTSHIP AND NESTING: The male has a complex courtship display. It is initiated with a pendulum display in which the bird shuttles back and forth in arcs—usually five to fifteen times—about one meter above the female, calling with a buzz throughout. At the end of each arc, the male spreads its tail and makes a rattling sound. This is often followed by a dive display: the male climbs to about twenty meters beyond the end of one of the pendulum arcs, then dives at full speed to the lowest point in the pendulum display. In pulling out of the dive, he makes a loud buzz, created by the tail feathers. The dive speed may reach over 100 km/hr (about 60 mph). The dive is also used to intimidate and drive off intruders. Nests are built about one-half to fifteen meters above ground, in dense brush vegetation in the fog zone of the Pacific Ocean coast. They are also built in eucalyptus (*Eucalyptus* sp.), pine (*Pinus* sp.), oak (*Quercus* sp.), and fir (*Abies* sp.) trees. They are made of pieces of leaves and grasses, with an inner layer of downy material as from willow (*Salix* sp.) catkins, all spun together with spider webbing. The sides of the nest are attached to sticks and branches adjacent to the main tree branch on which it rests. The outer layer of the nest is decorated with bits of lichen, moss, and bark. Females often construct new nests on top of old ones. The female usually lays two eggs forty-eight hours apart. Timing of incubation varies, with one report indicating that it begins after the laying of the second egg, and another indicating that it begins after the laying of the first egg. Reports also differ on the length of incubation, which may last from twelve to twenty-two days, a very broad range that requires more study. Fledging occurs twenty-one to twenty-five days after hatching. The female regurgitates her stomach contents by pumping it directly into the stomach of the nestling with her bill down its throat—the same as all hummingbirds.

NUTRITION AND MOLT: The Allen's Hummingbird depends on nectar from many species of flowering plants, and on small insects caught in the air or gleaned from foliage. Molt of body and flight feathers takes place from September through December.

ACCIDENTALS

Accidentals

GREEN VIOLET-EAR, GVIO (*Colibri thalassinus*) Colibrí Orejavioleta Verde (Sp.) *Colibri* = hummingbird; *thalassinus* = like a growing green branch

The Green Violet-ear is a hummingbird of central and southern Mexico. It ranges south into Central America. During different seasons, some individuals make the long trip north into the United States—where it has been seen in at least twelve states—and Canada. (It has been seen in two Canadian provinces.) The following sightings have been documented in eBird, an online international listing site available to the public, in recent years: La Crosse, Wisconsin, in October 1998; Mustang Island, Texas, in April 2002; Ontonagon, Michigan, in August 2002; Johnson City, Texas, in June 2003; Bastrop, Texas, in May 2007; and Cecil and Columbia, Maryland, in October 2011. Reports from birders indicate that there are many more recent sightings than are listed in eBird, especially in Texas.

IDENTIFICATION: The Green Violet-ear is a large, dark green bird almost the size of a Magnificent Hummingbird, with a medium-length dark bill that is slightly downcurved.

ADULT MALE: The green back may appear golden to dark green depending on the light angle. The undersides are green to blue green, with deeper blue to almost purple on the breast and paler green on the belly, fading to pale gray green toward the undertail coverts. Violet auricular patches are clearly visible. The tail is greenish blue above, especially on r1, and paler blue below. There is a wide dark blue band near the tips of all rectrices.

ADULT FEMALE AND JUVENILES: The plumage of adult females and juveniles of both sexes resembles the male plumage closely enough that Green Violet-ears are not likely to be confused with any other North American hummingbird. The plumage is paler than the male's, with less iridescence. The tail pattern is the same. Juveniles have pale buffy fringes to the crown and back feathers and corrugations on the maxilla, which disappear after fledging.

Green Violet-ear

SIMILAR SPECIES: The male Magnificent Hummingbird might appear similar to the Green Violet-ear, as both look very dark—almost black—when out of the sunlight. However, the Magnificent lacks the blue in the plumage of the Violet-ear and the dark band across the tail. The Magnificent has a purple crown, and the Violet-ear has purple auriculars. The Magnificent has a green gorget; the Violet-ear does not have a well-defined gorget.

Adult Green Violet-ear, showing the all-green plumage, slightly decurved black bill, blue tail, and dark blue-violet auriculars.

Adult, showing the bluish breast and violet auriculars.

Accidentals

Adult Green Violet-ear, dorsal view, showing the tail.

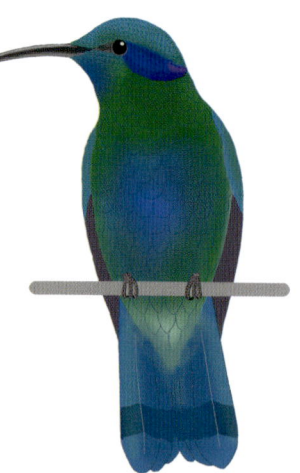

Adult, ventral view.

Tail, which is the same in both sexes and all ages.

GREEN-BREASTED MANGO, GREM (*Anthracothorax prevostii*) Mango Pechiverde (Sp.) *Anthraco* = coal or carbon (black); *thorax* = chest or breast; *prevostii* = commemorates the French naturalist Florent Prevost (Lesson)

The Green-breasted Mango is a tropical hummingbird whose principal range is in eastern Mexico south of Brownsville, Texas; on the Yucatán Peninsula; and south from there. It has been seen many times in southern Texas since 1988, and it has been documented according to eBird in Corpus Christi, Texas, in November 1997; Cabarros County, North Carolina, from November to December 2000; Hidalgo, Texas, from October 2004 to January 2005; Beloit, Wisconsin, in September 2007; Laurens, Georgia, from October 2007 to March 2008; and Greenwood, Louisiana, in August 2009. As not all birders report on eBird, there probably have been many more sightings.

IDENTIFICATION: The Green-breasted Mango is a large hummingbird with an all-black downcurved bill. Both sexes and all ages have a dark center stripe on the throat and breast and a brightly colored tail, unlike any other North American hummingbird.

ADULT MALE: The back is dark green, becoming more golden green on the rump and upper tail coverts. The gorget is iridescent green, with black stripe(s) reaching to the center of the breast. The bill is black and downcurved. The central tail feathers (r1) are darker coppery to greenish, depending on the light angle. R2 through r5 are mostly violet, with shades of red depending on the light angle. The vent band is white.

ADULT FEMALE: The back is not as dark green as the male's; it is more bronzy green. The undersides are white, except for the conspicuous black stripe that runs from the base of the mandible to the belly. Some adult females have a cinnamon streak from the chin partway down both sides of the central black streak. The flanks are greenish. The undertail coverts are greenish brown with white fringes. The tail is brightly patterned. The central pair of feathers (r1) is cinnamon to greenish brown. The other rectrices have white tips, a band of blue, and a band of violet up the feather. The tail is iridescent, and the violet band may show colors from red to brown to black, depending on the light angle.

JUVENILES: Juvenile males and females resemble the adult female, with a duller green to bronzy green back; white undersides; a black median stripe from breast to belly; and two cinnamon streaks from the chin, one on each side of the median black streak. Males may have some dark green feathers on the white undersides. The undertail coverts may be darker green than on the adult female. The juvenile female tail is not as brightly patterned as that of the juvenile male or adult female.

SIMILAR SPECIES: The female Mango, with its black median stripe on the undersides, is unique among North American hummingbirds. The male might be mistaken for a Magnificent Hummingbird, as both look black in poor light. The Magnificent has a straight bill, unlike the Mango, and the tails are very different when seen in good light: olive or bronzy green for the Magnificent and purple or lilac for the Mango.

Adult male Green-breasted Mango, dorsal view.

Adult male, ventral view.

Cinnamon Hummingbird

Adult female Green-breasted Mango, ventral view.

CINNAMON HUMMINGBIRD, CIHU (*Amazilia rutila*) Colibrí Canelo (Sp.) *Amazilia* = word of unknown origin; *rutila* = red or ruddy (DeLattre)

The Cinnamon Hummingbird is a very common resident of western coastal Mexico, from Sinaloa south and on the Yucatán Peninsula in the east. There are no recently documented records for this species in the United States on eBird, but the American Ornithologists' Union recognizes two records. One is of two birds in Patagonia, Arizona, in July 1992, and the other is of a single bird from Santa Teresa, New Mexico, in September 1993.

IDENTIFICATION: The Cinnamon Hummingbird is about the same size and shape as the other three *Amazilia* hummingbirds: Violet-crowned, Buff-bellied, and Berylline.

ADULTS: The sexes are alike. The back and crown are brilliant green in good sunlight. The upper tail coverts are rusty. The face, chin, throat, breast, and belly are pale to darker cinnamon or a rust color, leaving the vent band much lighter. The bill is red with a black tip. The notched tail is rusty, with dark brown to bronzy-green tips to all feathers.

Accidentals

JUVENILES: The sexes are alike. Juveniles resemble adults, except that the crowns and backs of very young birds have pale buffy tips to all feathers. The bill is mostly black. Corrugations on the maxilla remain for a short time postfledging.

SIMILAR SPECIES: From the back, the Cinnamon might be confused with a Buff-bellied Hummingbird. But when the bird turns toward you, you will see that the Cinnamon has an all-cinnamon or light rust-colored underside from chin to tail, unlike any of the other hummingbirds likely to be seen in North America.

Adult Cinnamon Hummingbird, dorsal view.

Adult, ventral view.

Adult in the field.

XANTUS'S HUMMINGBIRD, XAHU (*Hylocharis xantusii*)
Colibrí de Xantus (Sp.) *Hylocharis* = wood beauty; *xantusii* = named in honor of Hungarian naturalist John Xántus de Vesey (Lawrence)

Xantus's Hummingbird is a resident of the southern Baja Peninsula, Mexico. There are no records for this bird in northern Baja, and there are only two documented records north of Mexico: one in Ventura, California, in March 1988, and another on the Sunshine Coast, northwest of Vancouver, British Columbia, in November 1997. There is one other sight record from southern California.

IDENTIFICATION: Xantus's Hummingbird is a congener of the White-eared Hummingbird and has the same general appearance. It is a medium-sized hummingbird that appears bulky; it is often perched hunched forward. The head is dark with a conspicuous white postocular stripe. The tail is rusty.

ADULT MALE: The back is green; the head is dark blue black with a very conspicuous white postocular stripe. The gorget is dark green when seen in good light. The red bill has a black tip. The upper and undertail coverts are rusty. The undersides are rusty or buffy, with greenish speckling on the breast and some along the flanks. The vent band is white. The central tail feathers (r1) are green, with a rusty cast at the base. The other tail feathers (r2–r5) are rusty with faint greenish edges.

ADULT FEMALE: The green nape and back are duller than in the male. There is less blue black on the head. The bill is black above and dull red beneath, with an all-black tip. The undersides are buffy or rusty from chin to the undertail coverts, except for a lighter vent band. The central tail feathers (r1) are green, with rust on the basal portion. The other tail feathers (r2–r5) are rusty, with a black section that creates an irregular band just above the tail tips, which are lighter rust.

Accidentals

JUVENILES: Resemble adult females. There are pale buffy tips to the crown and upper back feathers and corrugations on the maxilla for a short time postfledging. Males are darker on the head, and they often have red at the base of the maxilla. Dark feathers are usually evident on the throat; these increase in number as the bird ages. The tail has some dark bronzy-green toward the tips of the four outer feathers (r2–r5). Juvenile females are paler than adult females, with more bronzy-green plumage on the head and back. The undersides are paler cinnamon, and the tail has more black in the subterminal band.

SIMILAR SPECIES: The Xantus's has very much the same size, shape, and general appearance as the White-eared Hummingbird. The main differences are the undersides of the Xantus's, which are rusty or buffy instead of white, and the tail, which is rusty instead of dark blue black.

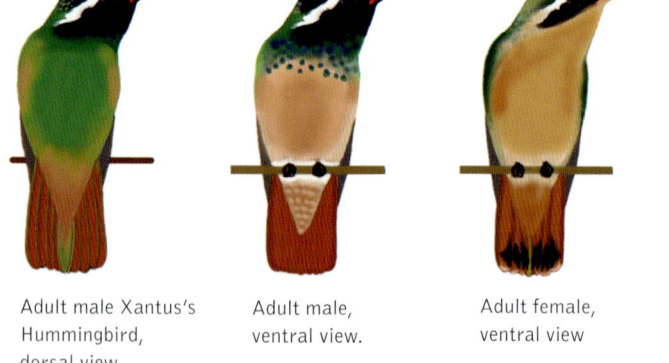

Adult male Xantus's Hummingbird, dorsal view.

Adult male, ventral view.

Adult female, ventral view

ANTILLEAN CRESTED HUMMINGBIRD, ACHU (*Orthorhyncus cristatus*) Colibrí Crestado (Sp.) *Ortho* = straight; *rhyncus* = bill; *cristatus* = crested (Linnaeus)

The Antillean Crested Hummingbird is resident in the Caribbean islands north to Puerto Rico. There are no documented records of the bird in the United States, although one was reported in Galveston, Texas, in 1967. The proximity of its range to Florida should encourage birders to watch for this species after a storm.

IDENTIFICATION: The Antillean Crested is a small hummingbird that is highly sexually dimorphic.

ADULT MALE: Unmistakable. The whole bird is small and very dark (almost black), with a bright blue-green to blue crest on the forehead. The bill is short, black, and straight. The back is dark to bronzy green, depending on the light angle. The undersides are sooty or charcoal and paler on the chin. The central tail feathers (r1) are greenish, and the outer four pairs (r2–r5) are blue black in good light.

ADULT FEMALE: The back and head are bright to olive green, depending on the light. There is a dusky mask back of the eye. There is no postocular white spot. The undersides are dull white, becoming grayer on the undertail coverts. The tail resembles that of the male except that r1 is more olive green, r2 through r5 are darker green at the base and dark blue distally, and r3 through r5 have white tips.

JUVENILES: Resemble adult females. The young male is darker than either the adult or juvenile female.

SIMILAR SPECIES: Females might be mistaken for female Ruby-throated or Black-chinned Hummingbirds, which have longer bills, thinner bodies, and a duskier or grayer breast and belly. No other North American hummingbird has a crest like the male of this species.

Accidentals

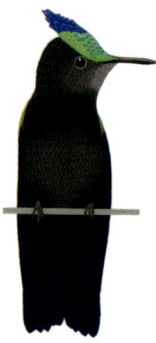

Adult male Antillean Crested Hummingbird, ventral view.

Adult female, ventral view.

Adult male tail, dorsal view.

Adult female tail, dorsal view.

CUBAN EMERALD, CUEM (*Chlorostilbon ricordii*) Zunzún (Sp.)
Chloro = green; *stilbon* = glittering; *ricordii* = in honor of Maryland doctor Alexandre Ricord (Gervais)

The Cuban Emerald is resident on the islands of Cuba, Bermuda, and the Bahamas. It has been reported over a dozen times on the Florida coast, but so far it has not been documented there by specimen or acceptable photograph. With its range so near to Florida, it is only a matter of time before its presence there is verified.

IDENTIFICATION: The Cuban Emerald is a small, dark hummingbird, with a medium-length bill and a very long, deeply forked tail.

ADULT MALE: The bird may appear all black out of sunlight. The head, back, and undersides are bright green, with shades of gold or blue depending on light angle. A dark mask extends to the auriculars. There is a conspicuous white postocular spot. The mandible is dull red, and the undertail coverts and vent band are white. The tail is very long with relatively wide feathers, and it is deeply forked. R1 and r2 are olive green; the others are dark blue to bluish green.

ADULT FEMALE: The crown and back are dull green, but in good light they are brighter. The face has a white posteriorly extended postocular stripe, which fades to gray and a dusky mask. The bill is black. The throat, breast, and belly are mostly dull white, and sometimes they are speckled with dull green and gray. The flanks are dull green to yellow green. The undertail coverts are white. The tail is long and deeply forked, with relatively wide individual feathers. The green upper tail coverts extend over green r1 and r2 feathers, while r3 through r5 are darker. Some birds have white tips on r4 and r5.

JUVENILES: Resemble adult females. Young females have shorter tails, with more white on the tips of r3 through r5. Juvenile males are darker and have longer, darker tails. They may have speckles of dark green feathers on the breast and belly.

Accidentals

SIMILAR SPECIES: No other North American hummingbird is all green with a long and deeply forked tail. Other than the tail, the male might be mistaken for a male Broad-billed Hummingbird, which has a shorter dark blue tail and a red bill with black tip. Although the female Broad-billed has a similar face pattern to the female Cuban Emerald, it lacks the long forked tail and has some pink or red on the underside of the mandible.

Adult male Cuban Emerald, ventral view.

Adult female, ventral view.

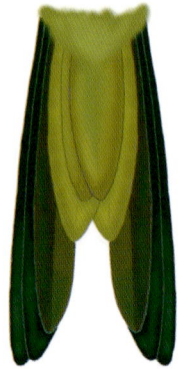

Adult male tail, dorsal view.

Adult female tail, dorsal view.

BAHAMA WOODSTAR, BAWO (*Calliphlox evelynae*) Colibrí de las Bahamas (Sp.) *Calli* = beautiful; *phlox* = flame; *evelynae* = named for Conrad Loddiges's daughter Evelyn, born in 1845 (Bourcier)

The Bahama Woodstar is resident on the Bahamas, south to the Turks and Caicos Islands. There are five records from Florida (in the Lantana, Homestead, and Miami areas) between 1961 and 1981, at least one with a specimen. No records are listed on eBird since 1981, but one Bahama Woodstar was reported from Lancaster County, Pennsylvania, in April 2013. Howell (2001) suggests that the species should be placed in the genus *Calothorax* (Sheartails), as it appears closer to the species in that genus—which includes the Lucifer Hummingbird—than it does to the Woodstars of tropical middle America.

IDENTIFICATION: Any small hummingbird with a long forked tail found in the eastern United States should be checked to see if it is this species.

ADULT MALE: The crown, back, and upper tail coverts are olive to golden green, depending on the angle of light. The face is dark with a small white postocular spot. The bill is short, straight, and black. The bright rose-red gorget appears black when the bird is not facing the sun. There is a distinct white collar under the gorget, which continues down the upper breast. The flanks and midbelly are dusky rufous mottled with green, becoming brighter rufous under the folded wings. The vent area is white. The undertail coverts are rufous. The tail is deeply forked, with r1 being very short, r2 of medium length, and r3 to r5 very long. The short central rectrices (r1) are olive green. R2 is mostly very dark, except for a green tip. R3 and r4 are rusty on the inner vanes and almost black on the outer vanes, and r5 is long, pointed, and all black.

ADULT FEMALE: The crown and back are dull olive to golden green, depending on the angle of light. There is a grayish mask in

Accidentals

the loral and auricular areas, as well as a white postocular spot. The throat is gray with some darker flecks. The breast is pale gray, leaving a white collar under the throat. The flanks are cinnamon or rusty, and they cover the midbelly. The cinnamon flanks are brightest under the folded wings; they are conspicuous in flight. The vent band is white, and the undertail coverts are rufous. The tail is long and notched. R1 is the shortest tail feather; it is green. R2 is longer and dark green to black at the tip, with some cinnamon at the base. R3 through r5 are longer and pale cinnamon at the tip. They have a subterminal black band and are green and cinnamon toward the base.

JUVENILES: Both sexes resemble adult females. The crown and back feathers have pale cinnamon tips, which wear off quickly after fledging. The young male has a cinnamon throat with occasional red spangles. The tail is longer and more forked than the adult female tail, and the cinnamon tips are smaller. The young female is very similar to the adult female after the scaly back has worn away, but there is less cinnamon at the base of the tail feathers.

SIMILAR SPECIES: Its red gorget gives the false impression that the male is just another male Ruby-throated Hummingbird, common in the eastern United States and especially in Florida, where the Bahama Woodstar is most likely to be found. But the Ruby-throated has a shorter forked black tail without any rufous. The Woodstar female's cinnamon belly, undertail coverts, and tail pattern are unlike those of any other North American hummingbird. Female and juvenile Rufous and Allen's Hummingbirds have rust on the flanks and in the tail, which is much brighter and deeper red brown than what is seen in the Woodstar female.

Bahama Woodstar

Adult male Bahama Woodstar, ventral view.

Adult female, ventral view.

Adult male tail, dorsal view.

Adult female tail, dorsal view.

Accidentals

BUMBLEBEE HUMMINGBIRD, BUHU (*Atthis heloisa*) Zumbador Mexicano (Sp.) *Atthis* = a name in Greek mythology; *heloisa* = a famous warrior, or may be from the Greek word for sun

The Bumblebee Hummingbird is a common resident of the Sierra Madre Occidental, from western Sinaloa south in Mexico. There is one record of two birds from Arizona's Huachuca Mountains in 1896; no other records or sightings of this species in North America exist.

IDENTIFICATION: The Bumblebee Hummingbird is very small, about the same size as or a little smaller than the Calliope Hummingbird. It acts more like a bee than a bird, moving slowly from flower to flower in a silent hovering flight, often almost on the ground. The black bill is short and straight.

ADULT MALE: The crown and back are olive to golden or bronzy green, with a faint cinnamon wash if seen in good light. There is a prominent white postocular stripe separating the green of the crown from the gorget. The gorget is magenta to rose to purple depending on light angle, with longer feathers lower on the throat. The undersides are white, with dull green to cinnamon flanks. The vent band is white, and the undertail coverts are pale cinnamon. The central rectrix (r1) is green. The other four pairs of rectrices have a rufous base and a wide black subterminal band. The tip of r2 is cinnamon, and the tips of r3 to r5 are white.

ADULT FEMALE: The crown and back are olive to bronzy or golden green. The face sometimes shows a very small postocular pale spot. The auriculars are dusky greenish gray. The throat is buffy, with many small spots. The breast and belly are washed with buff, brighter on the flanks, and almost white down the midcenter line. The undertail coverts are dull cinnamon. The tail pattern is the same as in the male, except that the basal rufous band is smaller and the black subterminal band is wider.

JUVENILES: Resemble adult females. The cinnamon or buffy tips to the back and crown feathers remain for a time (the length of

Bumblebee Hummingbird

which is unknown) after fledging. Corrugations on the maxilla are present for a short time postfledging. Young males have some green feathers or magenta spangles in the gorget. Young females have a paler plumage on the undersides. The tail feather tips may be buffy. More study of juvenile Bumblebees is needed.

SIMILAR SPECIES: The male Bumblebee's gorget is similar in color to the Calliope male's gorget, but the Bumblebee's iridescent gorget spangles cover the whole throat, while the Calliope's feathers are in distinct rows radiating down and out, leaving white spaces between rows of red feathers. The Bumblebee's tail has significantly more rufous at the base, the tip of r2 is buffy, and the tips of r3 to r5 are white in all ages and both sexes. Rufous and Allen's Hummingbirds are larger and have more rufous in the tail than Bumblebees. The Bumblebee's bill is much shorter. Rufous and Allen's are also much noisier and more aggressive at feeders.

Adult male Bumblebee Hummingbird, dorsal view.

Adult male, ventral view.

Adult female, dorsal view.

Adult female, ventral view.

MEASUREMENTS AND WEIGHTS OF
ADULT HUMMINGBIRDS

Measurements and Weights of Adult Hummingbirds

Species are listed in their currently accepted taxonomic order. Information was taken from a number of sources—very few sources for the accidental species, and several for the regularly occurring species. The bulk of the measurements of the latter species were obtained during HMN banding sessions in southeastern Arizona. There is a wide range in the body weights of some species. This can be due to fat accumulation prior to migration; the presence of an egg before laying; weight measurements that are taken at different times of day; variability among the scales used and their accuracy; and the natural variation among individuals.

Linear measurements also show wide ranges in some species. We did not measure total body length from bill tip to tail tip in any of our banding studies, nor did we measure tail lengths. These measurements are taken from the literature, where available. Wing chord variability can be due to some measurements of the wing taking place while it is flattened and straightened, whereas other measurements are of the length of the wing in a slightly curved position, as it naturally occurs. A measurement would be shortened if recorded when the bird is in molt and the longest primary has not grown in fully. Culmen lengths vary based on the positioning of the caliper or ruler at different places on the head. Some start where the last feather tip extends forward on the bill, and others push back to where the bill joins the skull. Tail length has the same problem, as it depends on how far you push the ruler into the rear end of the bird. Of course, birds are variable, and much of the variation recorded here is due to natural individual variation.

GREEN VIOLET-EAR

Body Length: Male & Female 107.0–114.0 mm
Wing Chord: Male 63.0–70.0 mm, Female 60.0–63.0 mm
Culmen: Male 18.0–22.0 mm, Female 19.0–22.0 mm
Weight: Male & Female 5.2–6.2 g

GREEN-BREASTED MANGO

Body Length: Male & Female 114.3–121.9 mm
Wing Chord: Male 63.0–69.0 mm, Female 62.0–69.0 mm
Culmen: Male 24.0–29.0 mm, Female 25.0–31.0 mm
Weight: Male & Female 5.5–6.5 g

ANTILLEAN CRESTED HUMMINGBIRD

Body Length: Male 78.7–95.3 mm, Female 73.7–78.7 mm
Wing Chord: no data
Culmen: no data
Weight: Male 2.2–4.3 g, Female 2.0–3.4 g

CUBAN EMERALD

Body Length: Male 91.4–101.6 mm, Female 96.5 mm
Wing Chord: Male 50.0–55.0 mm, Female 48.0–52.5 mm
Culmen: Male 14.5–18.5 mm, Female 17.5–19.0 mm
Weight: Male 2.9–4.6 g, Female 2.5–4.1 g

BROAD-BILLED HUMMINGBIRD

Body Length: Male & Female 88.9–101.6 mm
Wing Chord: Male 49.0–54.0 mm, Female 47.0–52.0 mm
Tail Length: Male 28.0–35.0 mm, Female 27.0–31.0 mm
Culmen: Male 20.0–22.0 mm, Female 20.0–23.0 mm
Weight: Male 2.7–4.5 g, Female 2.5–4.6 g

WHITE-EARED HUMMINGBIRD

Body Length: Male & Female 88.9–101.6 mm
Wing Chord: Male 51.0–59.0 mm, Female 49.0–55.0 mm
Tail Length: Male 31.0–37.0 mm, Female 29.0–33.0 mm
Culmen: Male 14.5–18.5 mm, Female 16.0–18.5 mm
Weight: Male Average 3.6 g, Female Average 3.2 g

XANTUS'S HUMMINGBIRD

Body Length: Male & Female 83.8–96.5 mm
Wing Chord: Male 49.0–54.0 mm, Female 43.0–53.0 mm
Culmen: Male 17.0–19.0 mm, Female 16.0–18.1 mm
Weight: Male 3.1–4.0 g, Female 3.1–3.9 g

BERYLLINE HUMMINGBIRD

Body Length: Male & Female 94.0–101.5 mm
Wing Chord: Male 49.0–58.0 mm, Female 50.0–56.0 mm
Tail Length: Male & Female 27.0–33.0 mm
Culmen: Male 18.0–21.0 mm, Female 19.0–21.0 mm
Weight: Male 4.4–5.7 g, Female 4.0–4.8 g

BUFF-BELLIED HUMMINGBIRD

Body Length: Male & Female 97.0–110.0 mm
Wing Chord: Male 45.0–65.0 mm, Female 51.0–59.0 mm
Tail Length: Male 33.0–40.0 mm, Female 33.0–36.0 mm
Culmen: Male 17.0–25.0 mm, Female 16.0–23.0 mm
Weight: Male 3.2–6.0 g, Female 2.0–7.0 g

CINNAMON HUMMINGBIRD

Body Length: Male & Female 101.6–114.3 mm
Wing Chord: Male 52.0–60.0 mm, Female 51.0–58.0 mm
Tail Length: Male & Female 31.0–37.0 mm
Culmen: Male 19.0–24.0 mm, Female 20.0–23.0 mm
Weight: Male & Female 4.0–5.2 g

VIOLET-CROWNED HUMMINGBIRD

Body Length: Male & Female 101.0–114.0 mm
Wing Chord: Male & Female 53.0–58.0 mm
Tail Length: Male & Female 26.0–33.0 mm
Culmen: Male & Female 20.5–25.9 mm
Weight: Male & Female 4.6–6.5 g

BLUE-THROATED HUMMINGBIRD

Body Length: Male & Female 121.9–134.6 mm
Wing Chord: Male 72.0–80.0 mm, Female 69.0–73.6 mm
Tail Length: Male 43.0–49.5 mm, Female 40.0–45.7 mm
Culmen: Male 22.0–24.7 mm, Female 24.5–27.5 mm
Weight: Male 7.6–8.8 g, Female 5.6–7.5 g

MAGNIFICENT HUMMINGBIRD

Body Length: Male & Female 119.4–134.5 mm
Wing Chord: Male 66.0–78.0 mm, Female 66.0–74.0 mm
Tail Length: Male 35.0–56.0 mm, Female 40.0–48.0 mm
Culmen: Male 25.0–32.5 mm, Female 25.0–33.0 mm
Weight: Male 6.8–9.6 g, Female 6.5–8.1 g

PLAIN-CAPPED STARTHROAT

Body Length: Male & Female 119.1–127.0 mm
Wing Chord: Male 63.0–71.0 mm, Female 64.0–69.0 mm
Culmen: Male 33.0–37.0 mm, Female 34.0–38.0 mm
Weight: Male 6.5–9.0 g, Female 6.8–7.8 g

BAHAMA WOODSTAR

Body Length: Male 80.0–95.0 mm, Female no data
Wing Chord: Male 37.0–41.0 mm, Female 41.0–46.0 mm
Culmen: Male 15.0–17.0 mm, Female 15.0–18.0 mm
Weight: Male & Female 2.5–3.5 g

LUCIFER HUMMINGBIRD

Body Length: Male 92.4–101.5 mm, Female 89.0–98.5 mm
Wing Chord: Male 36.0–41.0 mm, Female 39.0–44.0 mm
Tail Length: Male 27.0–32.0 mm, Female 22.0–27.0 mm
Culmen: Male 19.0–22.0 mm, Female 21.0–23.0 mm
Weight: Male 2.9–3.7 g, Female 3.1–3.9 g

RUBY-THROATED HUMMINGBIRD

Body Length: Male 81.0–91.0 mm, Female 88.0–95.0 mm
Wing Chord: Male 36.0–42.0 mm, Female 38.0–47.0 mm
Tail Length: Male 25.0–31.0 mm, Female 23.0–29.0 mm
Culmen: Male 14.0–20.0 mm, Female 16.0–21.0 mm
Weight: Male 2.3–5.2 g, Female 2.5–4.8 g

BLACK-CHINNED HUMMINGBIRD

Body Length: Male 83.8–91.4 mm, Female 88.9–96.5 mm
Wing Chord: Male 38.0–46.0 mm, Female 43.0–50.0 mm
Tail Length: Male 23.7–28.3 mm, Female 24.6–28.3 mm
Culmen: Male 16.0–20.6 mm, Female 17.9–22.1 mm
Weight: Male 2.7–4.0 g, Female 3.2–3.9 g

ANNA'S HUMMINGBIRD

Body Length: Male & Female 89.0–102.0 mm
Wing Chord: Male 46.2–51.3 mm, Female 46.0–52.0 mm
Tail Length: Male 29.1–33.3 mm, Female 25.0–31.0 mm
Culmen: Male 15.3–19.5 mm, Female 15.4–19.6 mm
Weight: Male 3.2–6.7 g, Female 3.0–5.7 g

COSTA'S HUMMINGBIRD

Body Length: Male 76.2–86.4 mm, Female 75.0–85.0 mm
Wing Chord: Male 42.3–45.7 mm, Female 42.9–46.5 mm
Tail Length: Male 21.0–24.7 mm, Female 21.3–25.3 mm
Culmen: Male 15.5–18.1 mm, Female 17.4–17.7 mm
Weight: Male 2.5–5.2 g, Female 2.5–3.5 g

CALLIOPE HUMMINGBIRD

Body Length: Male & Female 72.6–86.4 mm
Wing Chord: Male 37.2–41.3 mm, Female 40.1–44.0 mm
Tail Length: Male 19.6–22.1 mm, Female 19.7–23.3 mm
Culmen: Male 13.8–14.7 mm, Female 14.6–16.4 mm
Weight: Male 1.9–3.4 g, Female 2.2–3.2 g

BUMBLEBEE HUMMINGBIRD

Body Length: Male & Female 59.0–76.0 mm
Wing Chord: Male 32.0–38.0 mm, Female 35.0–38.0 mm
Tail Length: Male & Female 18.0–20.0 mm
Culmen: Male & Female 11.0–14.0 mm
Weight: Male 1.9–2.7 g, Female 2.0–2.6 g

BROAD-TAILED HUMMINGBIRD

Body Length: Male & Female 89.0–102.0 mm
Wing Chord: Male 46.3–51.0 mm, Female 47.9–52.1 mm
Tail Length: Male 28.7–35.4 mm, Female 26.8–32.6 mm
Culmen: Male 16.0–19.0 mm, Female 17.4–20.3 mm
Weight: Male 2.5–4.1 g, Female 3.1–4.4 g

RUFOUS HUMMINGBIRD

Body Length: Male & Female 81.3–94.1 mm
Wing Chord: Male 38.1–42.5 mm, Female 42.0–46.6 mm
Tail Length: Male 25.0–29.0 mm, Female 23.9–28.3 mm
Culmen: Male 14.4–18.0 mm, Female 16.4–19.0 mm
Weight: Male 2.0–4.0 g, Female 3.0–3.6 g
Width of r5: Male 1.8–2.6 mm, Female 2.7–4.0 mm

ALLEN'S HUMMINGBIRD

Body Length: Male & Female 81.0–89.0 mm
Wing Chord: Male 36.2–39.9 mm, Female 39.5–43.3 mm
Tail Length: Male 22.8–26.1 mm, Female 21.9–25.9 mm
Culmen: Male 14.0–17.0 mm, Female 15.8–18.8 mm
Weight: Male 2.5–3.9 g, Female 2.8–4.2 g
Width of r5: Male 1.1–1.9 mm, Female 1.8–2.8 mm

DIAGRAMS AND GLOSSARY OF HUMMINGBIRD CHARACTERISTICS AND TERMS

Diagrams and Glossary of Hummingbird Characteristics and Terms

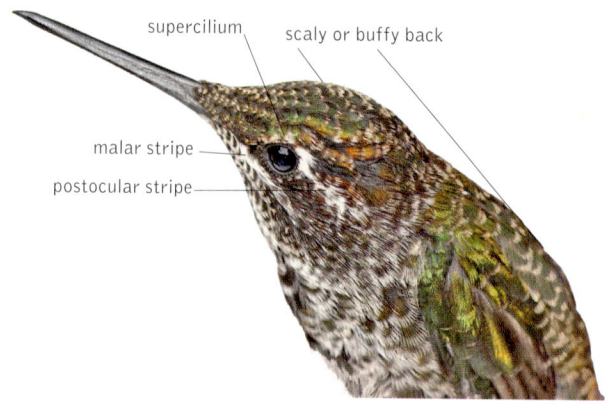

Juvenile Magnificent Hummingbird, showing parts and features mentioned in the species descriptions.

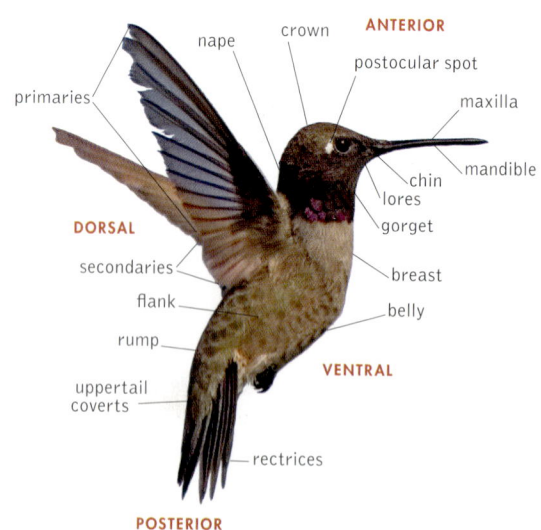

Adult male Black-chinned Hummingbird, showing parts and features mentioned in the species descriptions.

Diagrams and Glossary of Hummingbird Characteristics and Terms

ALTERNATE PLUMAGE: The plumage worn by adults of all species of hummingbird during the breeding season. See basic plumage.

ANTERIOR: Toward the front or head. On an opened wing, the anterior vane extends from the rachis forward in the outer primaries and laterally on the inner primaries.

AURICULARS: The feathers covering the ear opening behind the eye.

BARBS AND BARBULES: Barbs are the projections from each side of a feather's rachis (central shaft); these create the anterior and posterior vanes of the feather. Each barb has tiny projections on both sides, called barbules. The barbules hook together like Velcro to make the feather act as a flexible, integrated blade.

BASIC PLUMAGE: The plumage worn by adults of all species of hummingbirds during the nonbreeding season, usually during the winter in North America. See alternate plumage.

BILL CORRUGATIONS: Juvenile hummingbirds have thin grooves at about a forty-five-degree angle to the length of the bill, along the middle portion of the maxilla. These corrugations slowly disappear as the bill hardens, but they remain in most species from three to nine months after fledging. They serve to confirm that the bird is a juvenile.

BODY LENGTH: A linear measurement from the tip of the bill to the end of the tail with the bird lying flat, so that the bill, body, and tail are all on the same plane or in line. Body length is used in Measurements and Weights of Adult Hummingbirds.

CHIN: The area immediately below the base of the lower mandible.

COVERTS: Feathers that cover the base of flight feathers. For instance, undertail coverts are those feathers on the underside of the tail that cover the base of the rectrices. Primary and secondary coverts, uppertail coverts, and underwing coverts are other examples of this feather type.

CROWN: The top of the head.

CULMEN: The upper surface of the maxilla. The exposed culmen is used in Measurements and Weights of Adult Hummingbirds; this is the straight line distance from the tip of the bill to where the skin and feathers meet the upper surface of the bill. This distance is shorter than the actual culmen length when measured on a curved bill.

DORSAL: The back.

EARS: The ears are several long spangle feathers that project backward and sometimes outward from the rear of the gorget, as in Costa's and Anna's Hummingbirds. Obviously this term has nothing to do with hearing.

FEMORALS: Refers to the feathers on the upper leg, which are white and puffy on some hummingbirds.

FLANK: The side, the area of the body under and just below the folded flight feathers of a perched bird.

FLIGHT FEATHERS: See primary feather, secondary feather, and rectrix.

FORKED: Tail shape in which the outer feathers are much longer than the central tail feathers.

GORGET: The iridescent throat patch in many hummingbirds. The brilliant color of the gorget feathers is not due to pigment (although melanin, a dark pigment, is present in the feather) but is achieved through the refraction of light. Depending on the angle of the light and your position relative to the bird,

Diagrams and Glossary of Hummingbird Characteristics and Terms

sometimes the color is not seen, and the gorget appears black. In males of most species, the gorget feathers cover both the throat and the chin.

GRAVID: "Pregnant," applied to females carrying eggs shortly before laying, easily seen through the belly skin when the bird is in hand.

GROOVES: See bill corrugations.

LATERAL: Toward the side (as opposed to medial).

MALAR: The area of the face that extends from the bill back and down below the eye, along the jawline.

MANDIBLE: The lower bill.

MAXILLA: The upper bill.

MEDIAL: Toward the middle.

NARES: Nostrils.

NOTCHED: Tail shape in which the outer tail feathers (r5) are longer than the next pair (r4), and so on to the central pair (r1), which is shortest. The notch can be shallow or deep, depending on the relative length of the tail feathers. If the tail is very deeply notched, it is called "forked." The term "notched" is also used for adult male Rufous Hummingbirds, to describe a depression (or notch) on the medial side of the second rectrix (r2).

POSTERIOR: Toward the back or tail. On an opened wing, the posterior vane extends from the rachis backward in the outer primaries and medially in the inner primaries.

POSTOCULAR: Literally "behind the eye." Usually refers to the white spot or stripe behind the eye on most hummingbirds. The

Diagrams and Glossary of Hummingbird Characteristics and Terms

reason for the white spot is not known, but it may be meant to indicate to potential predators or rivals that the bird is watching them, while the real eye may be closed.

PRIMARY FEATHER: The outer ten flight feathers. Numbered p1 through p10, with p10 being the outermost.

RACHIS: The feather shaft.

RECTRIX (RECTRICES): Tail feathers, numbered r1 through r5. Rectrix 1 is in the center and r5 is the outermost feather on each side. All hummingbirds in this guide have ten tail feathers (rectrices).

RIPARIAN: In association with flowing water. Refers to a habitat along a river or stream, or in a valley with a seasonal creek, allowing plants that need wetter soil conditions to survive.

SCALY BACK: Also called "buff" or "buffy back," "scaly back" refers to the crown and dorsal body feathers of most juvenile hummingbirds, which are fringed at their tips with a pale shade—either whitish, grayish, cinnamon, or buffy—that contrasts with the darker shade of the rest of the feather. This results in the backs of most juveniles looking like scales, and it also serves to confirm that the bird is a juvenile. Most fringes wear off quickly, but some may remain through the winter.

SECONDARY FEATHER: Inner flight feather. The four to six flight feathers on the wing between the body and first primary. The secondary feathers are numbered inward (s1–s6) from the first primary (p1).

SHUTTLE DISPLAY: A courtship display in which the male hovers and moves rapidly from side to side while facing the female to show the maximum amount of color of his gorget plumage.

Diagrams and Glossary of Hummingbird Characteristics and Terms

SKY ISLAND: In the deserts of the southwestern United States, there are a number of isolated mountain ranges, which rise high above the desert floor (into the sky). The habitats of these islands are cooler and moister than the surrounding desert, and they attract populations of certain plants and animals. These include not only resident and migrating North American hummingbird species, but also hummingbirds whose primary habitat is in the upper elevations of the Sierra Madres of Mexico.

SPANGLE: A single iridescent gorget or crown feather.

SPATULATE: Feather shape in which the tip is wider than the base, as in the central pair of tail feathers of Calliope Hummingbirds.

SUPERCILIUM: The area from the bill extending posteriorly along the head, above the eye.

TAIL FEATHERS: See rectrix.

THROAT: The area between the chin and the neck. In hummingbirds, this area extends from the chin to the upper breast. Though generally synonymous with "gorget" in location, "throat" is usually used when there are no iridescent feathers (spangles).

TRAPLINE FEEDING: Some hummingbirds establish a pattern of feeding in which they fly a daily route from one flower patch or feeder to the next over an extended area—much as fur trappers did in earlier days, when checking their traps for animals.

VENT: The cloacal (anal) opening.

VENT BAND: The feathers, often white in hummingbirds, that arise at the base of the legs and around the cloacal opening.

VENTRAL: The undersides.

Diagrams and Glossary of Hummingbird Characteristics and Terms

WING CHORD: A measure of the folded wing, taken from the fold at the wrist to the end of the longest flight feather, often used to separate species and sexes. Wing chord is used in Measurements and Weights of Adult Hummingbirds.

XEROPHILE: One that loves the desert.

REFERENCES

References

American Ornithologists' Union. *Check-list of North American Birds.* 7th ed. Washington, DC: American Ornithologists' Union, 1998.

Baltosser, W. H. "Age and Sex Determination in the Calliope Hummingbird." *Western Birds* 25 (1994): 104–109.

———. "Age, Species, and Sex Determination of Four North American Hummingbirds." *North American Bird Bander* 12 (1987): 151–161.

Baltosser, W. H., and P. E. Scott. "Costa's Hummingbird (*Calypte costae*)." *The Birds of North America* 251 (1996), edited by A. Poole and F. Gill. Philadelphia, PA, and Washington, DC: The Academy of Natural Sciences and the American Ornithologists' Union.

Baltosser, W. H., and S. M. Russell. "Black-chinned Hummingbird (*Archilochus alexandri*)." *The Birds of North America* 495 (2000), edited by A. Poole and F. Gill. Philadelphia, PA, and Washington, DC: The Academy of Natural Sciences and the American Ornithologists' Union.

Calder, W. A. "Rufous Hummingbird (*Selasphorus rufus*)." *The Birds of North America* 53 (1993), edited by A. Poole and F. Gill. Philadelphia, PA, and Washington, DC: The Academy of Natural Sciences and the American Ornithologists' Union.

Calder, W. A., and L. L. Calder. "Broad-tailed Hummingbird (*Selasphorus platycercus*)." *The Birds of North America* 16 (1992), edited by A. Poole and F. Gill. Philadelphia, PA, and Washington, DC: The Academy of Natural Sciences and the American Ornithologists' Union.

———. "Calliope Hummingbird (*Stellula calliope*)." *The Birds of North America* 135 (1994), edited by A. Poole and F. Gill. Philadelphia, PA, and Washington, DC: The Academy of Natural Sciences and the American Ornithologists' Union.

Chavez-Ramirez, F., and A. Moreno-Valdez. "Buff-bellied Hummingbird (*Amazila yucatanensis*)." *The Birds of North America* 388 (1999), edited by A. Poole and F. Gill. Philadelphia, PA, and Washington, DC: The Academy of Natural Sciences and the American Ornithologists' Union.

Choate, E. A. *The Dictionary of American Bird Names.* Boston, MA: Harvard Common Press/Gambit, 1985.

Corman, T. E., and C. Wise-Gervais. *Arizona Breeding Bird Atlas.* Albuquerque: University of New Mexico Press, 2005.

Del Hoyo, J., A. Elliott, and J. Sargatal. *Barn-Owls to Hummingbirds.* Vol. 5 of *Handbook of the Birds of the World.* Rockville Center, NY: Lynx Edicions, 1999.

References

Dittman, D. L., and S. W. Cardiff. "The Alternate Plumage of the Ruby-throated Hummingbird." *Birding* 41 (2009): 32–35.

eBird. An online database of bird distribution and abundance [web application]. eBird, Ithaca, NY. http://www.ebird.org.

Fisher, D. *Early Southwest Ornithologists, 1528–1900*. Southwest Center Series. Tucson: University of Arizona Press, 2001.

Howell, C. A., and S. N. G. Howell. "Xantus's Hummingbird (*Hylocharis xantusii*)." *The Birds of North America* 554 (2000), edited by A. Poole and F. Gill. Philadelphia, PA, and Washington, DC: The Academy of Natural Sciences and the American Ornithologists' Union.

Howell, S. N. G. *Hummingbirds of North America: The Photographic Guide*. London and San Diego, CA: Academic Press, 2002.

Howell, S. N. G., and S. Webb. *A Guide to the Birds of Mexico and Northern Central America*. New York: Oxford University Press, 1995.

Hummingbird Monitoring Network. www.HumMonNet.org.

"Hummingbirds." Bird Calls and Songs. http://pjdeye.blogspot.com/2009/05/hummingbirds.html.

Johnsgard, P. A. *The Hummingbirds of North America*. 2nd ed. Washington, DC: Smithsonian Institution Press, 1997.

Kaufman, L. H. *Hummingbirds of the American West*. Tucson, Arizona: Rio Nuevo Publishers, 2001.

Keller, G. *Bird Songs of Southeastern Arizona and Sonora, Mexico*. Cornell Laboratory of Ornithology and Macaulay Library of Natural Sounds, 2001. CD. More information at http://www.artistdirect.com/nad/store/artist/album/0,,3139163,00.html#310hg62K5jFuzWUk.99.

Masear, T. *Fastest Things on Wings: Rescuing Hummingbirds in Hollywood*. Boston, MA: Houghton Mifflin Harcourt, 2015.

Mitchell, D. E. "Allen's Hummingbird (*Selasphorus sasin*)." *The Birds of North America* 501 (2000), edited by A. Poole and F. Gill. Philadelphia, PA, and Washington, DC: The Academy of Natural Sciences and the American Ornithologists' Union.

Powers, D. R. "Magnificent Hummingbird (*Eugenes fulgens*)." *The Birds of North America* 221 (1996), edited by A. Poole and F. Gill. Philadelphia, PA, and Washington, DC: The Academy of Natural Sciences and the American Ornithologists' Union.

Powers, D. R., and S. M. Wethington. "Broad-billed Hummingbird (*Cynanthus latirostris*)." *The Birds of North America* 430 (1999),

References

edited by A. Poole and F. Gill. Philadelphia, PA, and Washington, DC: The Academy of Natural Sciences and the American Ornithologists' Union.

Pyle, P. *Identification Guide to North American Birds*. Part 1. Bolinas, CA: Slate Creek Press, 1997.

Robinson, T. R., R. R. Sargent, and M. B. Sargent. "Ruby-throated Hummingbird (*Archilochus colubris*)." *The Birds of North America* 204 (1996), edited by A. Poole and F. Gill. Philadelphia, PA, and Washington, DC: The Academy of Natural Sciences and the American Ornithologists' Union.

Russell, S. M. "Anna's Hummingbird (*Calypte anna*)." *The Birds of North America* 226 (1996), edited by A. Poole and F. Gill. Philadelphia, PA, and Washington, DC: The Academy of Natural Sciences and the American Ornithologists' Union.

Russell, S. M., and R. O. Russell. *The North American Banders' Manual for Banding Hummingbirds*. Point Reyes Station, CA: The North American Banding Council, 2001. http://academic.keystone.edu/jskinner/OperationRubythroat/HUMM_MAN.pdf.

Sandrock, J., and J. C. Prior. *The Scientific Nomenclature of Birds in the Upper Midwest*. Iowa City: University of Iowa Press, 2014.

Scott, P. E. "Lucifer Hummingbird (*Calothorax lucifer*)." *The Birds of North America* 134 (1994), edited by A. Poole and F. Gill. Philadelphia, PA, and Washington, DC: The Academy of Natural Sciences and the American Ornithologists' Union.

Shackelford, C. F., M. M. Lindsay, and C. M. Klym. *Hummingbirds of Texas*. College Station: Texas A&M University Press, 2005.

Stiles, F. G. "Age and Sex Determination in Rufous and Allen's Hummingbirds." *Condor* 74 (1972): 25–32.

West, G. C., and C. A. Butler. *Do Hummingbirds Hum?* New Brunswick, NJ: Rutgers University Press, 2010.

Wethington, S. M. "Violet-crowned Hummingbird (*Amazilia violiceps*)." *The Birds of North America* 688 (2002), edited by A. Poole and F. Gill. Philadelphia, PA, and Washington, DC: The Academy of Natural Sciences and the American Ornithologists' Union.

Wethington, S. M., G. C. West, B. A. Carlson, N. L. Newfield, and S. J. Peters. "Longevity Records for North American Hummingbirds." *North American Bird Bander* 27 (2002): 131–133.

Williamson, S. L. "Blue-throated Hummingbird (*Lampornis clemenciae*)." *The Birds of North America* 531 (2001), edited by A. Poole and F. Gill. Philadelphia, PA, and Washington, DC: The Academy of Natural Sciences and the American Ornithologists' Union.

———. *Hummingbirds of North America*. A Peterson Field Guide. New York: Houghton Mifflin, 2001.

Xeno-Canto. Sharing Bird Sounds From Around the World. Founded in 2005 by Bob Planqué and Willem-Pier Vellinga of the Netherlands. http://www.xeno-canto.org/.

INDEX

Index

Allen's Hummingbird (*Selasphorus sasin*), xix, 150, 165, 167, 171, 174, 177, 179–188, 204, 207, 215

Amazilia, 34, 43, 56, 195

Anna's Hummingbird (*Calypte anna*), xvii, xviii, 113, 121–133, 134, 137, 138, 141, 142, 214, 219

Antillean Crested Hummingbird (*Orthorhynchus cristatus*), 199–200, 211

Archilochus, 98, 107, 113, 114, 117, 121, 128

Bahama Woodstar (*Calliphlox evelynae*), 203–205, 213

Berylline Hummingbird (*Amazilia beryllina*), xviii, 34–43, 54, 195, 212

Black-chinned Hummingbird (*Archilochus alexandrii*), xviii, 14, 30, 31, 55, 98, 104, 105, 106, 107–121, 132, 142, 199, 214, 224

Blue-throated Hummingbird (*Lampornis clemenciae*), 5, 6, 10, 11, 13, 14–24, 26, 43, 213

Broad-billed Hummingbird (*Cynanthus latirostris*), 76, 77–89, 202, 211

Broad-tailed Hummingbird (*Selasphorus platycercus*), xix, 14, 153–167, 177, 215

Buff-bellied Hummingbird (*Amazilia yucatanensis*), 42, 43–56, 195, 196, 212

Bumblebee Hummingbird (*Atthis heloisa*), 96, 152, 186, 206–207, 215

Calliope Hummingbird (*Selasphorus calliope*), 142, 144–153, 163, 206, 207, 214, 222

Calothorax, 89, 203

Calypte, 113, 121, 128, 134, 149

Cinnamon Hummingbird (*Amazilia rutila*), 42, 43, 54, 195–196, 212

Colibrí Barbinegro (Black-chinned Hummingbird), 107

Colibrí Canelo (Cinnamon Hummingbird), 195

Colibrí Corona-violeta (Violet-crowned Hummingbird), 56

Colibrí Crestado (Antillean Crested Hummingbird), 199

Colibrí de Anna (Anna's Hummingbird), 121

Colibrí de Berilo (Berylline Hummingbird), 34

Colibrí de Caliope (Calliope Hummingbird), 144

Colibrí de Costa (Costa's Hummingbird), 134

Colibrí de las Bahamas (Bahama Woodstar), 203

Colibrí de Xantus (Xantus's Hummingbird), 197

Colibrí Gorjirrubi (Ruby-throated Hummingbird), 98

Colibrí Magnifico (Magnificent Hummingbird), 2

Colibrí Orejavioleta Verde (Green Violet-ear), 190

Colibrí Orejiblanco (White-eared Hummingbird), 68

Colibrí Piquiancho (Broad-billed Hummingbird), 77

Colibrí Vientre-canelo (Buff-bellied Hummingbird), 43